Elliptic Partial Differential Equations
Second Edition

Qing Han
University of Notre Dame

Fanghua Lin
Courant Institute of Mathematical Sciences

1

Elliptic Partial Differential Equations

Second Edition

Courant Institute of Mathematical Sciences
New York University
New York, New York

American Mathematical Society
Providence, Rhode Island

2000 *Mathematics Subject Classification.* Primary 35–01, 35Jxx.

For additional information and updates on this book, visit
www.ams.org/bookpages/cln-1

Library of Congress Cataloging-in-Publication Data

Han, Qing.
Elliptic partial differential equations / Qing Han, Fanghua Lin. — 2nd ed.
p. cm. — (Courant lecture notes ; v. 1)
Includes bibliographical references.
ISBN 978-0-8218-5313-9 (alk. paper)
1. Differential equations, Elliptic. I. Lin, Fanghua. II. Title.

QA377.H3182 2011
515′.3533—dc22

2010051489

∞ The paper used in this book is acid-free and falls within the guidelines
established to ensure permanence and durability.
Visit the AMS home page at http://www.ams.org/

13 12 11 10 9 8 7 6 5 4 26 25 24 23 22 21

Dedicated to

Yansu, Raymond,
Wuxi, and Kathy

Contents

Preface

In the fall of 1992, the second author gave a course called "Intermediate PDEs" at the Courant Institute. The purpose of that course was to present some basic methods for obtaining various a priori estimates for second-order partial differential equations of elliptic type with particular emphasis on maximal principles, Harnack inequalities, and their applications. The equations one deals with are always linear, although they also obviously apply to nonlinear problems. Students with some knowledge of real variables and Sobolev functions should be able to follow the course without much difficulty.

In 1992, the lecture notes were taken by the first author. In 1995 at the University of Notre Dame, the first author gave a similar course. The original notes were then much extended, resulting in their present form.

It is not our intention to give a complete account of the related theory. Our goal is simply to provide these notes as a bridge between the elementary book of F. John [**9**], which also studies equations of other types, and the somewhat advanced book of D. Gilbarg and N. Trudinger [**8**], which gives a relatively complete account of the theory of elliptic equations of second order. We also hope our notes can serve as a bridge between the recent elementary book of N. Krylov [**11**] on the classical theory of elliptic equations developed before and around the 1960s and the book by Caffarelli and Cabré [**4**], which studies fully nonlinear elliptic equations, the theory obtained in the 1980s.

The authors wish to thank Karen Jacobs, Cheryl Huff, Joan Hoerstman, and Daisy Calderon for the wonderful typing job. The work was also supported by National Science Foundation Grants DMS No. 9401546 and DMS No. 9501122.

July 1997

In the new edition, we add a final chapter on the existence of solutions. In it we discuss several methods for proving the existence of solutions of primarily the Dirichlet problem for various types of elliptic equations. All these existence results are based on a priori estimates established in previous chapters.

December 2010

CHAPTER 1

Harmonic Functions

1.1. Guide

In this chapter we will use various methods to study harmonic functions. These include mean value properties, fundamental solutions, maximum principles, and energy methods. The four sections in this chapter are relatively independent of each other.

The materials in this chapter are rather elementary, but they contain several important ideas on the whole subject, and thus should be covered thoroughly. While doing Sections 1.2 and 1.3, the classic book by Protter and Weinberger [**13**] may be a very good reference. Also, when one reads Section 1.4, some statements concerning the Hopf maximal principle in Section 2.2 can be selected as exercises. The interior gradient estimates of Section 2.4 follow from the same arguments as those in the proof of Proposition 1.31 in Section 1.4.

1.2. Mean Value Properties

We begin this section with the definition of mean value properties. We assume that Ω is a connected domain in $\mathbb{R}^n$.

DEFINITION 1.1 For $u \in C(\Omega)$ we define

(i) u satisfies the first mean value property if

$$u(x) = \frac{1}{\omega_n r^{n-1}} \int\limits_{\partial B_r(x)} u(y)dS_y \quad \text{for any } B_r(x) \subset \Omega;$$

(ii) u satisfies the second mean value property if

$$u(x) = \frac{n}{\omega_n r^n} \int\limits_{B_r(x)} u(y)dy \quad \text{for any } B_r(x) \subset \Omega$$

where ω_n denotes the surface area of the unit sphere in $\mathbb{R}^n$.

REMARK 1.2. These two definitions are equivalent. In fact, if we write (i) as

$$u(x)r^{n-1} = \frac{1}{\omega_n} \int\limits_{\partial B_r(x)} u(y)dS_y,$$

we may integrate to get (ii). If we write (ii) as

$$u(x)r^n = \frac{n}{\omega_n} \int_{B_r(x)} u(y)dy,$$

we may differentiate to get (i).

REMARK 1.3. We may write the mean value properties in the following equivalent ways:

(i) u satisfies the first mean value property if

$$u(x) = \frac{1}{\omega_n} \int_{|w|=1} u(x + rw)dS_w \quad \text{for any } B_r(x) \subset \Omega;$$

(ii) u satisfies the second mean value property if

$$u(x) = \frac{n}{\omega_n} \int_{|z|\le 1} u(x + rz)dz \quad \text{for any } B_r(x) \subset \Omega.$$

Now we prove the maximum principle for the functions satisfying mean value properties.

PROPOSITION 1.4 *If $u \in C(\bar{\Omega})$ satisfies the mean value property in Ω, then u assumes its maximum and minimum only on $\partial\Omega$ unless u is constant.*

PROOF: We only prove for the maximum. Set

$$\Sigma = \{x \in \Omega : u(x) = M \equiv \max_{\bar{\Omega}} u\} \subset \Omega.$$

It is obvious that Σ is relatively closed. Next we show that Σ is open. For any $x_0 \in \Sigma$, take $\bar{B}_r(x_0) \subset \Omega$ for some $r > 0$. By the mean value property we have

$$M = u(x_0) = \frac{n}{\omega_n r^n} \int_{B_r(x_0)} u(y)dy \le M \frac{n}{\omega_n r^n} \int_{B_r(x_0)} dy = M.$$

This implies $u = M$ in $B_r(x_0)$. Hence Σ is both closed and open in Ω. Therefore either $\Sigma = \phi$ or $\Sigma = \Omega$. □

DEFINITION 1.5 A function $u \in C^2(\Omega)$ is harmonic if $\Delta u = 0$ in Ω.

THEOREM 1.6 *Let $u \in C^2(\Omega)$ be harmonic in Ω. Then u satisfies the mean value property in Ω.*

PROOF: Take any ball $B_r(x) \subset \Omega$. For $\rho \in (0, r)$, we apply the divergence theorem in $B_\rho(x)$ and get

$$(*) \quad \begin{aligned} \int_{B_\rho(x)} \Delta u(y)dy &= \int_{\partial B_\rho} \frac{\partial u}{\partial \nu} dS = \rho^{n-1} \int_{|w|=1} \frac{\partial u}{\partial \rho}(x + \rho w)dS_w \\ &= \rho^{n-1} \frac{\partial}{\partial \rho} \int_{|w|=1} u(x + \rho w)dS_w. \end{aligned}$$

Hence for harmonic function u we have for any $\rho \in (0, r)$

$$\frac{\partial}{\partial \rho} \int\limits_{|w|=1} u(x + \rho w) dS_w = 0.$$

Integrating from 0 to r we obtain

$$\int\limits_{|w|=1} u(x + rw) dS_w = \int\limits_{|w|=1} u(x) dS_w = u(x)\omega_n$$

or

$$u(x) = \frac{1}{\omega_n} \int\limits_{|w|=1} u(x + rw) dS_w = \frac{1}{\omega_n r^{n-1}} \int\limits_{\partial B_r(x)} u(y) dS_y.$$

□

REMARK 1.7. For a function u satisfying the mean value property, u is not required to be smooth. However a harmonic function is required to be C^2. We prove these two are equivalent.

THEOREM 1.8 *If $u \in C(\Omega)$ has mean value property in Ω, then u is smooth and harmonic in Ω.*

PROOF: Choose $\varphi \in C_0^\infty(B_1(0))$ with $\int_{B_1(0)} \varphi = 1$ and $\varphi(x) = \psi(|x|)$; i.e.,

$$\omega_n \int_0^1 r^{n-1} \psi(r) dr = 1.$$

We define $\varphi_\varepsilon(z) = \frac{1}{\varepsilon^n} \varphi(\frac{z}{\varepsilon})$ for $\varepsilon > 0$. Now for any $x \in \Omega$ consider $\varepsilon <$ dist $(x, \partial\Omega)$. Then we have

$$\begin{aligned}
\int\limits_{\Omega} u(y)\varphi_\varepsilon(y - x) dy &= \int u(x + y)\varphi_\varepsilon(y) dy \\
&= \frac{1}{\varepsilon^n} \int\limits_{|y|<\varepsilon} u(x + y)\varphi\left(\frac{y}{\varepsilon}\right) dy \\
&= \int\limits_{|y|<1} u(x + \varepsilon y)\varphi(y) dy \\
&= \int_0^1 r^{n-1} dr \int\limits_{\partial B_1(0)} u(x + \varepsilon rw)\varphi(rw) dS_w \\
&= \int_0^1 (r) r^{n-1} dr \int\limits_{|w|=1} u(x + \varepsilon rw) dS_w \\
&= u(x)\omega_n \int_0^1 (r) r^{n-1}\, dr = u(x)
\end{aligned}$$

where in the last equality we used the mean value property. Hence we get

$$u(x) = (\varphi_\varepsilon * u)(x) \quad \text{for any } x \in \Omega_\varepsilon = \{y \in \Omega;\ d(y, \partial\Omega) > \varepsilon\}.$$

Therefore u is smooth. Moreover, by formula $(*)$ in the proof of Theorem 1.2 and the mean value property we have

$$\begin{aligned}\int_{B_r(x)} \Delta u &= r^{n-1}\frac{\partial}{\partial r}\int_{|w|=1} u(x+rw)dS_w \\ &= r^{n-1}\frac{\partial}{\partial r}(\omega_n u(x)) = 0 \quad \text{for any } B_r(x) \subset \Omega.\end{aligned}$$

This implies $\Delta u = 0$ in Ω. □

REMARK 1.9. By combining Theorem 1.6 and Theorem 1.8, we conclude that harmonic functions are smooth and satisfy the mean value property. Hence harmonic functions satisfy the maximum principle, a consequence of which is the uniqueness of solution to the following Dirichlet problem in a bounded domain

$$\begin{aligned}\Delta u &= f \quad \text{in } \Omega,\\ u &= \varphi \quad \text{on } \partial\Omega,\end{aligned}$$

for $f \in C(\Omega)$ and $\varphi \in C(\partial\Omega)$. In general uniqueness does not hold for an unbounded domain. Consider the following Dirichlet problem in the unbounded domain Ω

$$\begin{aligned}\Delta u &= 0 \quad \text{in } \Omega,\\ u &= 0 \quad \text{on } \partial\Omega.\end{aligned}$$

First consider the case $\Omega = \{x \in \mathbb{R}^n; |x| > 1\}$. For $n = 2, u(x) = \log|x|$ is a solution. Note $u \to \infty$ as $r \to \infty$. For $n \geq 3$, $u(x) = |x|^{2-n} - 1$ is a solution. Note $u \to -1$ as $r \to \infty$. Hence u is bounded. Next, consider the upper half space $\Omega = \{x \in \mathbb{R}^n; x_n > 0\}$. Then $u(x) = x_n$ is a nontrivial solution, which is unbounded.

In the following we discuss the gradient estimates.

LEMMA 1.10 *Suppose $u \in C(\bar{B}_R)$ is harmonic in $B_R = B_R(x_0)$. Then there holds*

$$|Du(x_0)| \leq \frac{n}{R}\max_{\bar{B}_R}|u|.$$

PROOF: For simplicity we assume $u \in C^1(\bar{B}_R)$. Since u is smooth, then $\Delta(D_{x_i}u) = 0$, that is, $D_{x_i}u$ is also harmonic in B_R. Hence $D_{x_i}u$ satisfies the mean value property. By the divergence theorem we have

$$D_{x_i}u(x_0) = \frac{n}{\omega_n R^n}\int_{B_R(x_0)} D_{x_i}u(y)dy = \frac{n}{\omega_n R^n}\int_{\partial B_R(x_0)} u(y)\nu_i\, dS_y,$$

which implies

$$|D_{x_i}u(x_0)| \leq \frac{n}{\omega_n R^n}\max_{\partial B_R}|u| \cdot \omega_n R^{n-1} \leq \frac{n}{R}\max_{\bar{B}_R}|u|.$$

□

LEMMA 1.11 *Suppose $u \in C(\bar{B}_R)$ is a nonnegative harmonic function in $B_R = B_R(x_0)$. Then there holds*

$$|Du(x_0)| \leq \frac{n}{R}u(x_0).$$

PROOF: As before by the divergence theorem and the nonnegativeness of u we have

$$|D_{x_i}u(x_0)| \leq \frac{n}{\omega_n R^n} \int_{\partial B_R(x_0)} u(y)dS_y = \frac{n}{R}u(x_0)$$

where in the last equality we used the mean value property. □

COROLLARY 1.12 *A harmonic function in $\mathbb{R}^n$ bounded from above or below is constant.*

PROOF: Suppose u is a harmonic function in $\mathbb{R}^n$. We will prove that u is a constant if $u \geq 0$. In fact, for any $x \in \mathbb{R}^n$ we apply Lemma 1.11 to u in $B_R(x)$ and then let $R \to \infty$. We conclude that $Du(x) = 0$ for any $x \in \mathbb{R}^n$. □

PROPOSITION 1.13 *Suppose $u \in C(\bar{B}_R)$ is harmonic in $B_R = B_R(x_0)$. Then there holds for any multi-index α with $|\alpha| = m$*

$$|D^\alpha u(x_0)| \leq \frac{n^m e^{m-1} m!}{R^m} \max_{\bar{B}_R} |u|.$$

PROOF: We prove by induction. It is true for $m = 1$ by Lemma 1.10. Assume it holds for m. Consider $m+1$. For $0 < \theta < 1$, define $r = (1-\theta)R \in (0, R)$. We apply Lemma 1.10 to u in B_r and get

$$|D^{m+1}u(x_0)| \leq \frac{n}{r} \max_{\bar{B}_r} |D^m u|.$$

By the induction assumption we have

$$\max_{\bar{B}_r} |D^m u| \leq \frac{n^m \cdot e^{m-1} \cdot m!}{(R-r)^m} \max_{\bar{B}_R} |u|.$$

Hence we obtain

$$|D^{m+1}u(x_0)| \leq \frac{n}{r} \cdot \frac{n^m e^{m-1} m!}{(R-r)^m} \max_{B_R} |u| = \frac{n^{m+1} e^{m-1} m!}{R^{m+1}\theta^m(1-\theta)} \max_{\bar{B}_R} |u|.$$

Take $\theta = \frac{m}{m+1}$. This implies

$$\frac{1}{\theta^m(1-\theta)} = \left(1 + \frac{1}{m}\right)^m (m+1) < e(m+1).$$

Hence the result is established for any single derivative. For any multi-index $\alpha = (\alpha_1, \ldots, \alpha_n)$ we have

$$\alpha_1! \cdots \alpha_n! \leq (|\alpha|)!.$$

□

THEOREM 1.14 *Harmonic function is analytic.*

PROOF: Suppose u is a harmonic function in Ω. For fixed $x \in \Omega$, take $B_{2R}(x) \subset \Omega$ and $h \in \mathbb{R}^n$ with $|h| \leq R$. We have by Taylor expansion

$$u(x+h) = u(x) + \sum_{i=1}^{m-1} \frac{1}{i!}\left[\left(h_1 \frac{\partial}{\partial x_1} + \cdots + h_n \frac{\partial}{\partial x_n}\right)^i u\right](x) + R_m(h)$$

where

$$R_m(h) = \frac{1}{m!}\left[\left(h_1 \frac{\partial}{\partial x_1} + \cdots + h_n \frac{\partial}{\partial x_n}\right)^m u\right](x_1 + \theta h_1, \ldots, x_n + \theta h_n)$$

for some $\theta \in (0,1)$. Note $x + h \in B_R(x)$ for $|h| < R$. Hence by Proposition 1.13 we obtain

$$|R_m(h)| \leq \frac{1}{m!}|h|^m \cdot n^m \cdot \frac{n^m e^{m-1} m!}{R^m} \max_{\bar{B}_{2R}} |u| \leq \left(\frac{|h| n^2 e}{R}\right)^m \max_{\bar{B}_{2R}} |u|.$$

Then for any h with $|h|n^2 e < \frac{R}{2}$ there holds $R_m(h) \to 0$ as $m \to \infty$. □

Next we prove the Harnack inequality.

THEOREM 1.15 *Suppose u is harmonic in Ω. Then for any compact subset K of Ω there exists a positive constant $C = (\Omega, K)$ such that if $u \geq 0$ in Ω, then*

$$\frac{1}{C} u(y) \leq u(x) \leq C u(y) \quad \textit{for any } x, y \in K.$$

PROOF: By mean value property, we can prove if $B_{4R}(x_0) \subset \Omega$, then

$$\frac{1}{c} u(y) \leq u(x) \leq c u(y) \quad \text{for any } x, y \in B_R(x_0)$$

where c is a positive constant depending only on n. Now for the given compact subset K, take $x_1, \ldots, x_N \in K$ such that $\{B_R(x_i)\}$ covers K with $4R < \operatorname{dist}(K, \partial\Omega)$. Then we can choose $C = c^N$. □

We finish this section by proving a result, originally due to Weyl. Suppose u is harmonic in Ω. Then integrating by parts we have

$$\int_\Omega u \Delta\varphi = 0 \quad \text{for any } \varphi \in C_0^2(\Omega).$$

The converse is also true.

THEOREM 1.16 *Suppose $u \in C(\Omega)$ satisfies*

$$\int_\Omega u \Delta\varphi = 0 \quad \textit{for any } \varphi \in C_0^2(\Omega). \tag{1.1}$$

Then u is harmonic in Ω.

PROOF: We claim for any $B_r(x) \subset \Omega$ there holds

$$r \int_{\partial B_r(x)} u(y)\, dS_y = n \int_{B_r(x)} u(y)\, dy. \tag{1.2}$$

Then we have

$$\begin{aligned}
&\frac{d}{dr}\left(\frac{1}{\omega_n r^{n-1}}\int_{\partial B_r(x)} u(y)dS_y\right)\\
&= \frac{n}{\omega_n}\frac{d}{dr}\left(\frac{1}{r^n}\int_{B_r(x)} u(y)dy\right)\\
&= \frac{n}{\omega_n}\left\{-\frac{n}{r^{n+1}}\int_{B_r(x)} u(y)dy + \frac{1}{r^n}\int_{\partial B_r(x)} u(y)dS_y\right\} = 0.
\end{aligned}$$

This implies

$$\frac{1}{\omega_n r^{n-1}}\int_{\partial B_r(x)} u(y)dS_y = \text{const.}$$

This constant is $u(x)$ if we let $r \to 0$. Hence we have

$$u(x) = \frac{1}{\omega_n r^{n-1}}\int_{\partial B_r(x)} u(y)dS_y \quad \text{for any } B_r(x) \subset \Omega.$$

Next we prove (1.2) for $n \geq 3$. For simplicity we assume that $x = 0$. Set

$$\varphi(y,r) = \begin{cases} (|y|^2 - r^2)^n, & |y| \leq r, \\ 0, & |y| > r, \end{cases}$$

and then $\varphi_k(y,r) = (|y|^2 - r^2)^{n-k}(2(n-k+1)|y|^2 + n(|y|^2 - r^2))$ for $|y| \leq r$ and $k = 2, 3, \ldots, n$. Direct calculation shows $\varphi(\cdot, r) \in C_0^2(\Omega)$ and

$$\Delta_y \varphi(y,r) = \begin{cases} 2n\varphi_2(y,r), & |y| \leq r, \\ 0, & |y| > r. \end{cases}$$

By assumption (1.1) we have

$$\int_{B_r(0)} u(y)\varphi_2(y,r)dy = 0.$$

Now we prove if for some $k = 2, 3, \ldots, n-1$,

$$\int_{B_r(0)} u(y)\varphi_k(y,r)dy = 0, \tag{1.3}$$

then

$$\int_{B_r(0)} u(y)\varphi_{k+1}(y,r)dy = 0. \tag{1.4}$$

In fact, we differentiate (1.3) with respect to r and get

$$\int_{\partial B_r(0)} u(y)\varphi_k(y,r)dy + \int_{B_r(0)} u(y)\frac{\partial \varphi_k}{\partial r}(y,r)dy = 0.$$

For $2 \le k < n$, $\varphi_k(y,r) = 0$ for $|y| = r$. Then we have

$$\int_{B_r(0)} u(y)\frac{\partial \varphi_k}{\partial r}(y,r)dy = 0.$$

Direct calculation yields $\frac{\partial \varphi_k}{\partial r}(y,r) = (-2r)(n-k+1)\varphi_{k+1}(y,r)$. Hence we have (1.4). Therefore by taking $k = n-1$ in (1.4) we conclude

$$\int_{B_r(0)} u(y)((n+2)|y|^2 - nr^2)dy = 0.$$

Differentiating with respect to r again we get (1.2). □

1.3. Fundamental Solutions

We begin this section by seeking a harmonic function u, that is, $\Delta u = 0$ in $\mathbb{R}^n$, which depends only on $r = |x-a|$ for some fixed $a \in \mathbb{R}^n$. We set $v(r) = u(x)$. This implies

$$v'' + \frac{n-1}{r}v' = 0$$

and hence

$$v(r) = \begin{cases} c_1 + c_2 \log r, & n = 2, \\ c_3 + c_4 r^{2-n}, & n \ge 3, \end{cases}$$

where c_i are constants for $i = 1, 2, 3, 4$. We are interested in a function with a singularity such that

$$\int_{\partial B_r} \frac{\partial v}{\partial r} dS = 1 \quad \text{for any } r > 0.$$

Hence we set for any fixed $a \in \mathbb{R}^n$

$$\Gamma(a,x) = \frac{1}{2\pi}\log|a-x| \qquad \text{for } n = 2$$

$$\Gamma(a,x) = \frac{1}{\omega_n(2-n)}|a-x|^{2-n} \quad \text{for } n \ge 3.$$

To summarize, we have that for fixed $a \in \mathbb{R}^n$, $\Gamma(a,x)$ is harmonic at $x \ne a$, that is,

$$\Delta_x \Gamma(a,x) = 0 \quad \text{for any } x \ne a$$

and has a singularity at $x = a$. Moreover, it satisfies

$$\int_{\partial B_r(a)} \frac{\partial \Gamma}{\partial n_x}(a,x)dS_x = 1 \quad \text{for any } r > 0.$$

Now we prove the Green's identity.

THEOREM 1.17 *Suppose Ω is a bounded domain in $\mathbb{R}^n$ and that $u \in C^1(\bar{\Omega}) \cap C^2(\Omega)$. Then for any $a \in \Omega$ there holds*

$$u(a) = \int_{\Omega} \Gamma(a,x)\Delta u(x)dx - \int_{\partial\Omega} \left(\Gamma(a,x)\frac{\partial u}{\partial n_x}(x) - u(x)\frac{\partial \Gamma}{\partial n_x}(a,x) \right) dS_x.$$

REMARK 1.18.

(i) For any $a \in \Omega$, $\Gamma(a, \cdot)$ is integrable in Ω although it has a singularity.
(ii) For $a \notin \bar{\Omega}$, the expression in the right side gives 0.
(iii) By letting $u = 1$ we have $\int_{\partial\Omega} \frac{\partial \Gamma}{\partial n_x}(a,x)dS_x = 1$ for any $a \in \Omega$.

PROOF: We apply Green's formula to u and $\Gamma(a, \cdot)$ in the domain $\Omega \setminus B_r(a)$ for small $r > 0$ and get

$$\int_{\Omega \setminus B_r(a)} (\Gamma \Delta u - u \Delta \Gamma)dx =$$

$$\int_{\partial\Omega} \left(\Gamma \frac{\partial u}{\partial n} - u \frac{\partial \Gamma}{\partial n} \right) dS_x - \int_{\partial B_r(a)} \left(\Gamma \frac{\partial u}{\partial n} - u \frac{\partial \Gamma}{\partial n} \right) dS_x.$$

Note $\Delta\Gamma = 0$ in $\Omega \setminus B_r(a)$. Then we have

$$\int_{\Omega} \Gamma \Delta u \, dx = \int_{\partial\Omega} \left(\Gamma \frac{\partial u}{\partial n} - u \frac{\partial \Gamma}{\partial n} \right) dS_x - \lim_{r \to 0} \int_{\partial B_r(a)} \left(\Gamma \frac{\partial u}{\partial n} - u \frac{\partial \Gamma}{\partial n} \right) dS_x.$$

For $n \geq 3$, we get by definition of Γ

$$\left| \int_{\partial B_r(a)} \Gamma \frac{\partial u}{\partial n} dS \right| = \left| \frac{1}{(2-n)\omega_n} r^{2-n} \int_{\partial B_r(a)} \frac{\partial u}{\partial n} dS \right|$$
$$\leq \frac{r}{n-2} \sup_{\partial B_r(a)} |Du| \to 0 \qquad \text{as } r \to 0,$$

$$\int_{\partial B_r(a)} u \frac{\partial \Gamma}{\partial n} dS = \frac{1}{\omega_n r^{n-1}} \int_{\partial B_r(a)} u \, dS \to u(a) \qquad \text{as } r \to 0.$$

We get the same conclusion for $n = 2$ in the same way. □

REMARK 1.19. We may employ the local version of the Green's identity to get gradient estimates without using the mean value property. Suppose $u \in C(\bar{B}_1)$ is harmonic in B_1. For any fixed $0 < r < R < 1$ choose a cutoff function $\varphi \in C_0^\infty(B_R)$ such that $\varphi = 1$ in B_r and $0 \leq \varphi \leq 1$. Apply the Green's formula to u and $\varphi\Gamma(a, \cdot)$ in $B_1 \setminus B_\rho(a)$ for $a \in B_r$ and ρ small enough. We proceed as in the proof of Theorem 1.17 and obtain

$$u(a) = -\int_{r<|x|<R} u(x)\Delta_x(\varphi(x)\Gamma(a,x))dx \quad \text{for any } a \in B_r(0).$$

Hence one may prove (without using the mean value property)

$$\sup_{B_{1/2}} |u| \le c\Big(\int_{B_1} |u|^p\Big)^{1/p} \quad \text{and} \quad \sup_{B_{1/2}} |Du| \le c \max_{B_1} |u|$$

where c is a constant depending only on n.

Now we begin to discuss the Green's functions. Suppose Ω is a bounded domain in $\mathbb{R}^n$. Let $u \in C^1(\bar{\Omega}) \cap C^2(\Omega)$. We have by Theorem 1.17 for any $x \in \Omega$

$$u(x) = \int_{\Omega} \Gamma(x,y)\Delta u(y)dy - \int_{\partial\Omega} \Big(\Gamma(x,y)\frac{\partial u}{\partial n_y}(y) - u(y)\frac{\partial \Gamma}{\partial n_y}(x,y)\Big)dS_y.$$

If u solves the Dirichlet boundary value problem

$$(*) \qquad \begin{cases} \Delta u = f & \text{in } \Omega, \\ u = \varphi & \text{on } \partial\Omega, \end{cases}$$

for some $f \in C(\bar{\Omega})$ and $\varphi \in C(\partial\Omega)$, then u can be expressed in terms of f and φ, with one *unknown term.* We want to eliminate this term by adjusting Γ.

For any fixed $x \in \Omega$, consider

$$\gamma(x,y) = \Gamma(x,y) + \Phi(x,y)$$

for some $\Phi(x,\cdot) \in C^2(\bar{\Omega})$ with $\Delta_y \Phi(x,y) = 0$ in Ω. Then Theorem 1.17 can be expressed as follows for any $x \in \Omega$

$$u(x) = \int_{\Omega} \gamma(x,y)\Delta u(y)dy - \int_{\partial\Omega} \Big(\gamma(x,y)\frac{\partial u}{\partial n_y}(y) - u(y)\frac{\partial \gamma}{\partial n_y}(x,y)\Big)dS_y$$

since the extra $\Phi(x,\cdot)$ is harmonic. Now by choosing Φ appropriately, we are led to the important concept of Green's function.

For each fixed $x \in \Omega$ choose $\Phi(x,\cdot) \in C^1(\bar{\Omega}) \cap C^2(\Omega)$ such that

$$\begin{cases} \Delta_y \Phi(x,y) = 0 & \text{for } y \in \Omega, \\ \Phi(x,y) = -\Gamma(x,y) & \text{for } y \in \partial\Omega. \end{cases}$$

Denote the resulting $\gamma(x,y)$ by $G(x,y)$, which is called Green's function. If such a G exists, then the solution u to the Dirichlet problem $(*)$ can be expressed as

$$u(x) = \int_{\Omega} G(x,y)f(y)dy + \int_{\partial\Omega} \varphi(y)\frac{\partial G}{\partial n_y}(x,y)dS_y.$$

Note that Green's function $G(x,y)$ is defined as a function of $y \in \bar{\Omega}$ for each fixed $x \in \Omega$.

Now we discuss some properties of G as a function of x and y. Our first observation is that the Green's function is unique. This is proved by the maximum principle since the difference of two Green's functions are harmonic in Ω with zero boundary value. In fact, we have more.

PROPOSITION 1.20 *Green's function $G(x, y)$ is symmetric in $\Omega \times \Omega$; that is, $G(x, y) = G(y, x)$ for $x \neq y \in \Omega$.*

PROOF: Pick $x_1, x_2 \in \Omega$ with $x_1 \neq x_2$. Choose $r > 0$ small such that $B_r(x_1) \cap B_r(x_2) = \varnothing$. Set $G_1(y) = G(x_1, y)$ and $G_2(y) = G(x_2, y)$. We apply Green's formula in $\Omega \setminus B_r(x_1) \cup B_r(x_2)$ and get

$$\int_{\Omega \setminus B_r(x_1) \cup B_r(x_2)} (G_1 \Delta G_2 - G_2 \Delta G_1) = \int_{\partial\Omega} \left(G_1 \frac{\partial G_2}{\partial n} - G_2 \frac{\partial G_1}{\partial n} \right) dS$$
$$- \int_{\partial B_r(x_1)} \left(G_1 \frac{\partial G_2}{\partial n} - G_2 \frac{\partial G_1}{\partial n} \right) dS$$
$$- \int_{\partial B_r(x_2)} \left(G_1 \frac{\partial G_2}{\partial n} - G_2 \frac{\partial G_1}{\partial n} \right) dS.$$

Since G_i is harmonic for $y \neq x_i$, $i = 1, 2$, and vanishes on $\partial\Omega$, we have

$$\int_{\partial B_r(x_1)} \left(G_1 \frac{\partial G_2}{\partial n} - G_2 \frac{\partial G_1}{\partial n} \right) dS + \int_{\partial B_r(x_2)} \left(G_1 \frac{\partial G_2}{\partial n} - G_2 \frac{\partial G_1}{\partial n} \right) dS = 0.$$

Note the left side has the same limit as the left side in the following as $r \to 0$:

$$\int_{\partial B_r(x_1)} \left(\Gamma \frac{\partial G_2}{\partial n} - G_2 \frac{\partial \Gamma}{\partial n} \right) dS + \int_{\partial B_r(x_2)} \left(G_1 \frac{\partial \Gamma}{\partial n} - \Gamma \frac{\partial G_1}{\partial n} \right) dS = 0.$$

Since

$$\int_{\partial B_r(x_1)} \Gamma \frac{\partial G_2}{\partial n} dS \to 0, \qquad \int_{\partial B_r(x_2)} \Gamma \frac{\partial G_1}{\partial n} dS \to 0 \quad \text{as } r \to 0,$$

$$\int_{\partial B_r(x_1)} G_2 \frac{\partial \Gamma}{\partial n} dS \to G_2(x_1), \qquad \int_{\partial B_r(x_2)} G_1 \frac{\partial \Gamma}{\partial n} dS \to G_1(x_2) \quad \text{as } r \to 0,$$

we obtain $G_2(x_1) - G_1(x_2) = 0$ or equivalently $G(x_2, x_1) = G(x_1, x_2)$. □

PROPOSITION 1.21 *There holds for $x, y \in \Omega$ with $x \neq y$*

$$0 > G(x, y) > \Gamma(x, y) \qquad \textit{for } n \geq 3$$
$$0 > G(x, y) > \Gamma(x, y) - \frac{1}{2\pi} \log \operatorname{diam}(\Omega) \quad \textit{for } n = 2.$$

PROOF: Fix $x \in \Omega$ and write $G(y) = G(x, y)$. Since $\lim_{y \to x} G(y) = -\infty$ then there exists an $r > 0$ such that $G(y) < 0$ in $B_r(x)$. Note that G is harmonic in $\Omega \setminus B_r(x)$ with $G = 0$ on $\partial\Omega$ and $G < 0$ on $\partial B_r(x)$. The maximum principle implies $G(y) < 0$ in $\Omega \setminus B_r(x)$ for such $r > 0$. Next, consider the other part of the inequality. Recall the definition of the Green's function

$$G(x, y) = \Gamma(x, y) + \Phi(x, y) \quad \text{where} \begin{cases} \Delta\Phi = 0 & \text{in } \Omega, \\ \Phi = -\Gamma & \text{on } \partial\Omega. \end{cases}$$

For $n \geq 3$, we have

$$\Gamma(x, y) = \frac{1}{(2-n)\omega_n}|x-y|^{2-n} < 0 \quad \text{for } y \in \partial\Omega,$$

which implies $\Phi(x, \cdot) > 0$ on $\partial\Omega$. By the maximum principle, we have $\Phi > 0$ in Ω. For $n = 2$ we have

$$\Gamma(x, y) = \frac{1}{2\pi}\log|x-y| \leq \frac{1}{2\pi}\log \operatorname{diam}(\Omega) \quad \text{for } y \in \partial\Omega.$$

Hence the maximum principle implies $\Phi > -\frac{1}{2\pi}\log \operatorname{diam}(\Omega)$ in Ω. □

We may calculate the Green's functions for some special domains.

PROPOSITION 1.22 *The Green's function for the ball $B_R(0)$ is given by*

(i) $$G(x, y) = \frac{1}{(2-n)\omega_n}\left(|x-y|^{2-n} - \left|\frac{R}{|x|}x - \frac{|x|}{R}y\right|^{2-n}\right) \quad \textit{for } n \geq 3,$$

(ii) $$G(x, y) = \frac{1}{2\pi}\left(\log|x-y| - \log\left|\frac{R}{|x|}x - \frac{|x|}{R}y\right|\right) \quad \textit{for } n = 2.$$

PROOF: Fix $x \neq 0$ with $|x| < R$. Consider $X \in \mathbb{R}^n \setminus \bar{B}_R$ with X the multiple of x and $|X| \cdot |x| = R^2$, that is, $X = \frac{R^2}{|x|^2}x$. In other words, X and x are reflexive with respect to the sphere ∂B_R. Note the map $x \longmapsto X$ is conformal; that is, it preserves angles. If $|y| = R$, we have by similarity of triangles

$$\frac{|x|}{R} = \frac{R}{|X|} = \frac{|y-x|}{|y-X|},$$

which implies

$$|y-x| = \frac{|x|}{R}|y-X| = \left|\frac{|x|}{R}y - \frac{R}{|x|}x\right| \quad \text{for any } y \in \partial B_R. \tag{1.5}$$

Therefore, in order to have zero boundary value, we take for $n \geq 3$

$$G(x, y) = \frac{1}{(2-n)\omega_n}\left(\frac{1}{|x-y|^{n-2}} - \left(\frac{R}{|x|}\right)^{n-2}\frac{1}{|y-X|^{n-2}}\right).$$

The case $n = 2$ is similar. □

Next, we calculate the normal derivative of Green's function on the sphere.

COROLLARY 1.23 *Suppose G is the Green's function in $B_R(0)$. Then there holds*

$$\frac{\partial G}{\partial n}(x, y) = \frac{R^2 - |x|^2}{\omega_n R|x-y|^n} \quad \textit{for any } x \in B_R \textit{ and } y \in \partial B_R.$$

PROOF: We just consider the case $n \geq 3$. Recall with $X = R^2x/|x|^2$

$$G(x, y) = \frac{1}{(2-n)\omega_n}\left(|x-y|^{2-n} - \left(\frac{R}{|x|}\right)^{n-2}|y-X|^{2-n}\right)$$
$$\text{for } x \in B_R, y \in \partial B_R.$$

Hence we have for such x and y

$$D_{y_i}G(x,y) = -\frac{1}{\omega_n}\left(\frac{x_i - y_i}{|x-y|^n} - \left(\frac{R}{|x|}\right)^{n-2} \cdot \frac{X_i - y_i}{|X-y|^n}\right) = \frac{y_i}{\omega_n R^2}\frac{R^2 - |x|^2}{|x-y|^n}$$

by (1.5) in the proof of Proposition 1.22. We obtain with $n_i = \frac{y_i}{R}$ for $|y| = R$

$$\frac{\partial G}{\partial n}(x,y) = \sum_{i=1}^{n} n_i D_{y_i} G(x,y) = \frac{1}{w_n R} \cdot \frac{R^2 - |x|^2}{|x-y|^n}.$$

□

Denote by $K(x,y)$ the function in Corollary 1.23 for $x \in \Omega, y \in \partial\Omega$. It is called a Poisson kernel and has the following properties:

(i) $K(x,y)$ is smooth for $x \neq y$;
(ii) $K(x,y) > 0$ for $|x| < R$;
(iii) $\int_{|y|=R} K(x,y)dS_y = 1$ for any $|x| < R$.

The following result gives the existence of harmonic functions in balls with prescribed Dirichlet boundary value.

THEOREM 1.24 (Poisson Integral Formula) *For $\varphi \in C(\partial B_R(0))$, the function u defined by*

$$u(x) = \begin{cases} \int_{\partial B_R(0)} K(x,y)\varphi(y)dS_y, & |x| < R, \\ \varphi(x), & |x| = R, \end{cases}$$

satisfies $u \in C(\bar{\Omega}) \cap C^\infty(\Omega)$ and

$$\begin{cases} \Delta u = 0 & \text{in } \Omega, \\ u = \varphi & \text{on } \partial\Omega. \end{cases}$$

For the proof, see [**9**, pp. 107–108].

REMARK 1.25. In the Poisson integral formula, by letting $x = 0$, we have

$$u(0) = \frac{1}{\omega_n R^{n-1}} \int_{\partial B_R(0)} \varphi(y)dS_y,$$

which is the mean value property.

LEMMA 1.26 (Harnack's Inequality) *Suppose u is harmonic in $B_R(x_0)$ and $u \geq 0$. Then there holds*

$$\left(\frac{R}{R+r}\right)^{n-2}\frac{R-r}{R+r}u(x_0) \leq u(x) \leq \left(\frac{R}{R-r}\right)^{n-2}\frac{R+r}{R-r}u(x_0)$$

where $r = |x - x_0| < R$.

PROOF: We may assume $x_0 = 0$ and $u \in C(\bar{B}_R)$. Note that u is given by the Poisson integral formula

$$u(x) = \frac{1}{\omega_n R}\int_{\partial B_R} \frac{R^2 - |x|^2}{|y-x|^n}u(y)dS_y.$$

Since $R - |x| \leq |y - x| \leq R + |x|$ for $|y| = R$, we have

$$\frac{1}{\omega_n R} \cdot \frac{R - |x|}{R + |x|} \left(\frac{1}{R + |x|} \right)^{n-2} \int_{\partial B_R} u(y) dS_y \leq u(x) \leq \frac{1}{\omega_n R} \cdot \frac{R + |x|}{R - |x|} \left(\frac{1}{R - |x|} \right)^{n-2} \int_{\partial B_R} u(y) dS_y.$$

The mean value property implies

$$u(0) = \frac{1}{\omega_n R^{n-1}} \int_{\partial B_R} u(y) dS_y.$$

This finishes the proof. □

COROLLARY 1.27 *If harmonic function u in $\mathbb{R}^n$ is bounded above or below, then* $u \equiv$ const.

PROOF: We assume $u \geq 0$ in $\mathbb{R}^n$. Take any point $x \in \mathbb{R}^n$ and apply Lemma 1.26 to any ball $B_R(0)$ with $R > |x|$. We obtain

$$\left(\frac{R}{R + |x|} \right)^{n-2} \frac{R - |x|}{R + |x|} u(0) \leq u(x) \leq \left(\frac{R}{R - |x|} \right)^{n-2} \frac{R + |x|}{R - |x|} u(0),$$

which implies $u(x) = u(0)$ by letting $R \to +\infty$. □

Next we prove a result concerning the removable singularity.

THEOREM 1.28 *Suppose u is harmonic in $B_R \setminus \{0\}$ and satisfies*

$$u(x) = \begin{cases} o(\log |x|), & n = 2, \\ o(|x|^{2-n}), & n \geq 3, \end{cases} \quad \text{as } |x| \to 0.$$

Then u can be defined at 0 so that it is C^2 and harmonic in B_R.

PROOF: Assume u is continuous in $0 < |x| \leq R$. Let v solve

$$\begin{cases} \Delta v = 0 & \text{in } B_R, \\ v = u & \text{on } \partial B_R. \end{cases}$$

We will prove $u = v$ in $B_R \setminus \{0\}$. Set $w = v - u$ in $B_R \setminus \{0\}$ and $M_r = \max_{\partial B_r} |w|$. We prove for $n \geq 3$. It is obvious that

$$|w(x)| \leq M_r \cdot \frac{r^{n-2}}{|x|^{n-2}} \quad \text{on } \partial B_r.$$

Note w and $\frac{1}{|x|^{n-2}}$ are harmonic in $B_R \setminus B_r$. Hence the maximum principle implies

$$|w(x)| \leq M_r \cdot \frac{r^{n-2}}{|x|^{n-2}} \quad \text{for any } x \in B_R \setminus B_r$$

where $M_r = \max_{\partial B_r} |v - u| \le \max_{\partial B_r} |v| + \max_{\partial B_r} |u| \le M + \max_{\partial B_r} |u|$ with $M = \max_{\partial B_R} |u|$. Hence we have for each fixed $x \neq 0$

$$|w(x)| \le \frac{r^{n-2}}{|x|^{n-2}} M + \frac{1}{|x|^{n-2}} r^{n-2} \max_{\partial B_r} |u| \to 0 \quad \text{as } r \to 0,$$

that is $w = 0$ in $B_R \setminus \{0\}$. □

1.4. Maximum Principles

In this section we will use the maximum principle to derive the interior gradient estimate and the Harnack inequality.

THEOREM 1.29 *Suppose $u \in C^2(B_1) \cap C(\bar{B}_1)$ is a subharmonic function in B_1; that is, $\Delta u \ge 0$. Then there holds*

$$\sup_{B_1} u \le \sup_{\partial B_1} u.$$

PROOF: For $\varepsilon > 0$ we consider $u_\varepsilon(x) = u(x) + \varepsilon |x|^2$ in B_1. Then simple calculation yields

$$\Delta u_\varepsilon = \Delta u + 2n\varepsilon \ge 2n\varepsilon > 0.$$

It is easy to see, by a contradiction argument, that u_ε cannot have an interior maximum, in particular,

$$\sup_{B_1} u_\varepsilon \le \sup_{\partial B_1} u_\varepsilon.$$

Therefore we have

$$\sup_{B_1} u \le \sup_{B_1} u_\varepsilon \le \sup_{\partial B_1} u + \varepsilon.$$

We finish the proof by letting $\varepsilon \to 0$. □

REMARK 1.30. The result still holds if B_1 is replaced by any bounded domain.

The next result is the interior gradient estimate for harmonic functions. The method is due to Bernstein back in 1910.

PROPOSITION 1.31 *Suppose u is harmonic in B_1. Then there holds*

$$\sup_{B_{1/2}} |Du| \le c \sup_{\partial B_1} |u|$$

where $c = c(n)$ is a positive constant. In particular, for any $\alpha \in [0, 1]$ there holds

$$|u(x) - u(y)| \le c|x - y|^\alpha \sup_{\partial B_1} |u| \quad \textit{for any } x, y \in B_{1/2}$$

where $c = c(n, \alpha)$ is a positive constant.

PROOF: Direct calculation shows that

$$\Delta(|Du|^2) = 2 \sum_{i,j=1}^{n} (D_{ij} u)^2 + 2 \sum_{i=1}^{n} D_i u D_i(\Delta u) = 2 \sum_{i,j=1}^{n} (D_{ij} u)^2$$

where we used $\Delta u = 0$ in B_1. Hence $|Du|^2$ is a subharmonic function. To get interior estimates we need a cutoff function. For any $\varphi \in C_0^1(B_1)$ we have

$$\Delta(\varphi|Du|^2) = (\Delta\varphi)|Du|^2 + 4\sum_{i,j=1}^{n} D_i\varphi D_j u D_{ij}u + 2\varphi\sum_{i,j=1}^{n}(D_{ij}u)^2.$$

By taking $\varphi = \eta^2$ for some $\eta \in C_0^1(B_1)$ with $\eta \equiv 1$ in $B_{1/2}$, we obtain by the Hölder inequality

$$\begin{aligned}\Delta(\eta^2|Du|^2) &= 2\eta\Delta\eta|Du|^2 + 2|D\eta|^2|Du|^2\\ &\quad + 8\eta\sum_{i,j=1}^{n} D_i\eta D_j u D_{ij}u + 2\eta^2\sum_{i,j=1}^{n}(D_{ij}u)^2\\ &\geq (2\eta\Delta\eta - 6|D\eta|^2)|Du|^2 \geq -C|Du|^2\end{aligned}$$

where C is a positive constant depending only on η. Note that $\Delta(u^2) = 2|Du|^2 + 2u\Delta u = 2|Du|^2$ since u is harmonic. By taking α large enough we get

$$\Delta(\eta^2|Du|^2 + \alpha u^2) \geq 0.$$

We may apply Theorem 1.29 (the maximum principle) to get the result. □

Next we derive the Harnack inequality.

LEMMA 1.32 *Suppose u is a nonnegative harmonic function in B_1. Then there holds*

$$\sup_{B_{1/2}} |D\log u| \leq C$$

where $C = C(n)$ is a positive constant.

PROOF: We may assume $u > 0$ in B_1. Set $v = \log u$. Then direct calculation shows

$$\Delta v = -|Dv|^2.$$

We need the interior gradient estimate on v. Set $w = |Dv|^2$. Then we get

$$\Delta w + 2\sum_{i=1}^{n} D_i v D_i w = 2\sum_{i,j=1}^{n}(D_{ij}v)^2.$$

As before we need a cutoff function. First note

$$\sum_{i,j=1}^{n}(D_{ij}v)^2 \geq \sum_{i}^{n}(D_{ii}v)^2 \geq \frac{1}{n}(\Delta v)^2 = \frac{|Dv|^4}{n} = \frac{w^2}{n}. \tag{1.6}$$

Take a nonnegative function $\varphi \in C_0^1(B_1)$. We obtain by the Hölder inequality

$$\begin{aligned}
&\Delta(\varphi w) + 2\sum_{i=1}^n D_i v D_i(\varphi w) \\
&= 2\varphi \sum_{i,j=1}^n (D_{ij}v)^2 + 4\sum_{i,j=1}^n D_i\varphi D_j v D_{ij} v + 2w\sum_{i=1}^n D_i\varphi D_i v + (\Delta\varphi)w \\
&\geq \varphi \sum_{i,j=1}^n (D_{ij}v)^2 - 2|D\varphi||Dv|^3 - \left(|\Delta\varphi| + C\frac{|D\varphi|^2}{\varphi}\right)|Dv|^2
\end{aligned}$$

if φ is chosen such that $|D\varphi|^2/\varphi$ is bounded in B_1. Choose $\varphi = \eta^4$ for some $\eta \in C_0^1(B_1)$. Hence for such fixed η we obtain by (1.1)

$$\begin{aligned}
&\Delta(\eta^4 w) + 2\sum_{i=1}^n D_i v D_i(\eta^4 w) \\
&\geq \frac{1}{n}\eta^4|Dv|^4 - C\eta^3|D\eta||Dv|^3 - 4\eta^2(\eta\Delta\eta + C|D\eta|^2)|Dv|^2 \\
&\geq \frac{1}{n}\eta^4|Dv|^4 - C\eta^3|Dv|^3 - C\eta^2|Dv|^2
\end{aligned}$$

where C is a positive constant depending only on n and η. Hence we get by the Hölder inequality

$$\Delta(\eta^4 w) + 2\sum_{i=1}^n D_i v D_i(\eta^4 w) \geq \frac{1}{n}\eta^4 w^2 - C$$

where C is a positive constant depending only on n and η.

Suppose $\eta^4 w$ attains its maximum at $x_0 \in B_1$. Then $D(\eta^4 w) = 0$ and $\Delta(\eta^4 w) \leq 0$ at x_0. Hence there holds

$$\eta^4 w^2(x_0) \leq C(n, \eta).$$

If $w(x_0) \geq 1$, then $\eta^4 w(x_0) \leq C(n)$. Otherwise $\eta^4 w(x_0) \leq w(x_0) \leq \eta^4(x_0)$. In both cases we conclude

$$\eta^4 w \leq C(n, \eta) \quad \text{in } B_1.$$

□

COROLLARY 1.33 *Suppose u is a nonnegative harmonic function in B_1. Then there holds*

$$u(x_1) \leq Cu(x_2) \quad \textit{for any } x_1, x_2 \in B_{1/2}$$

where C is a positive constant depending only on n.

PROOF: We may assume $u > 0$ in B_1. For any $x_1, x_2 \in B_{1/2}$ by simple integration we obtain with Lemma 1.32

$$\log\frac{u(x_1)}{u(x_2)} \leq |x_1 - x_2|\int_0^1 |D\log u(tx_2 + (1-t)x_1)|dt \leq C|x_1 - x_2|.$$

□

Next we prove a quantitative Hopf lemma.

PROPOSITION 1.34 *Suppose $u \in C(\bar{B}_1)$ is a harmonic function in $B_1 = B_1(0)$. If $u(x) < u(x_0)$ for any $x \in \bar{B}_1$ and some $x_0 \in \partial B_1$, then there holds*

$$\frac{\partial u}{\partial n}(x_0) \geq C(u(x_0) - u(0))$$

where C is a positive constant depending only on n.

PROOF: Consider a positive function v in B_1 defined by

$$v(x) = e^{-\alpha|x|^2} - e^{-\alpha}.$$

It is easy to see

$$\Delta v(x) = e^{-\alpha|x|^2}(-2\alpha n + 4\alpha^2|x|^2) > 0 \quad \text{for any } |x| \geq \tfrac{1}{2}$$

if $\alpha \geq 2n + 1$. Hence for such fixed α the function v is subharmonic in the region $A = B_1 \setminus B_{1/2}$. Now define for $\varepsilon > 0$

$$h_\varepsilon(x) = u(x) - u(x_0) + \varepsilon v(x).$$

This is also a subharmonic function, that is, $\Delta h_\varepsilon \geq 0$ in A. Obviously $h_\varepsilon \leq 0$ on ∂B_1 and $h_\varepsilon(x_0) = 0$. Since $u(x) < u(x_0)$ for $|x| = \frac{1}{2}$ we may take $\varepsilon > 0$ small such that $h_\varepsilon(x) < 0$ for $|x| = \frac{1}{2}$. Therefore by Theorem 1.29 h_ε assumes at the point x_0 its maximum in A. This implies

$$\frac{\partial h_\varepsilon}{\partial n}(x_0) \geq 0 \quad \text{or} \quad \frac{\partial u}{\partial n}(x_0) \geq -\varepsilon\frac{\partial v}{\partial n}(x_0) = 2\alpha\varepsilon e^{-\alpha} > 0.$$

Note that so far we have only used the subharmonicity of u. We estimate ε as follows. Set $w(x) = u(x_0) - u(x) > 0$ in B_1. Obviously w is a harmonic function in B_1. By Corollary 1.33 (the Harnack inequality) there holds

$$\inf_{B_{1/2}} w \geq c(n)w(0) \quad \text{or} \quad \max_{B_{1/2}} u \leq u(x_0) - c(n)(u(x_0) - u(0)).$$

Hence we may take

$$\varepsilon = \delta c(n)(u(x_0) - u(0))$$

for δ small, depending on n. This finishes the proof. □

To finish this section we prove a global Hölder continuity result.

LEMMA 1.35 *Suppose $u \in C(\bar{B}_1)$ is a harmonic function in B_1 with $u = \varphi$ on ∂B_1. If $\varphi \in C^\alpha(\partial B_1)$ for some $\alpha \in (0, 1)$, then $u \in C^{\alpha/2}(\bar{B}_1)$. Moreover, there holds*

$$\|u\|_{C^{\alpha/2}(\bar{B}_1)} \leq C\|\varphi\|_{C^\alpha(\partial B_1)}$$

where C is a positive constant depending only on n and α.

PROOF: First the maximum principle implies that $\inf_{\partial B_1} \varphi \leq u \leq \sup_{\partial B_1} \varphi$ in B_1. Next we claim that for any $x_0 \in \partial B_1$ there holds

$$\sup_{x \in B_1} \frac{|u(x) - u(x_0)|}{|x - x_0|^{\alpha/2}} \leq 2^{\alpha/2} \sup_{x \in \partial B_1} \frac{|\varphi(x) - \varphi(x_0)|}{|x - x_0|^\alpha}. \tag{1.7}$$

Lemma 1.35 follows easily from (1.7). For any $x, y \in B_1$, set $d_x = \mathrm{dist}(x, \partial B_1)$ and $d_y = \mathrm{dist}(y, \partial B_1)$. Suppose $d_y \le d_x$. Take $x_0, y_0 \in \partial B_1$ such that $|x - x_0| = d_x$ and $|y - y_0| = d_y$. Assume first that $|x - y| \le d_x/2$. Then $y \in \overline{B}_{d_x/2}(x) \subset B_{d_x}(x) \subset B_1$. We apply Proposition 1.31 (scaled version) to $u - u(x_0)$ in $B_{d_x}(x)$ and get by (1.7)

$$d_x^{\alpha/2}\frac{|u(x) - u(y)|}{|x - y|^{\alpha/2}} \le C|u - u(x_0)|_{L^\infty(B_{d_x}(x))} \le C d_x^{\alpha/2}\|\varphi\|_{C^\alpha(\partial B_1)}.$$

Hence we obtain

$$|u(x) - u(y)| \le C|x - y|^{\alpha/2}\|\varphi\|_{C^\alpha(\partial B_1)}.$$

Assume now that $d_y \le d_x \le 2|x - y|$. Then by (1.7) again we have

$$\begin{aligned}|u(x) - u(y)| &\le |u(x) - u(x_0)| + |u(x_0) - u(y_0)| + |u(y_0) - u(y)|\\ &\le C(d_x^{\alpha/2} + |x_0 - y_0|^{\alpha/2} + d_y^{\alpha/2})\|\varphi\|_{C^\alpha(\partial B_1)}\\ &\le C|x - y|^{\alpha/2}\|\varphi\|_{C^\alpha(\partial B_1)}\end{aligned}$$

since $|x_0 - y_0| \le d_x + |x - y| + d_y \le 5|x - y|$.

In order to prove (1.7) we assume $B_1 = B_1((1, 0, \ldots, 0))$, $x_0 = 0$, and $\varphi(0) = 0$. Define $K = \sup_{x\in\partial B_1} |\varphi(x)|/|x|^\alpha$. Note $|x|^2 = 2x_1$ for $x \in \partial B_1$. Therefore for $x \in \partial B_1$ there holds

$$\varphi(x) \le K|x|^\alpha \le 2^{\alpha/2}Kx_1^{\alpha/2}.$$

Define $v(x) = 2^{\alpha/2}Kx_1^{\alpha/2}$ in B_1. Then we have

$$\Delta v(x) = 2^{\alpha/2}K \cdot \frac{\alpha}{2}\left(\frac{\alpha}{2} - 1\right)x_1^{\alpha/2-2} < 0 \quad \text{in } B_1.$$

Theorem 3.1 implies

$$u(x) \le v(x) = 2^{\alpha/2}Kx_1^{\alpha/2} \le 2^{\alpha/2}K|x|^{\alpha/2} \quad \text{for any } x \in B_1.$$

Considering $-u$ similarly, we get

$$|u(x)| \le 2^{\alpha/2}K|x|^{\alpha/2} \quad \text{for any } x \in B_1.$$

This proves (1.7). □

1.5. Energy Method

In this section we discuss harmonic functions by using the energy method. In general we assume throughout this section that $a_{ij} \in C(B_1)$ satisfies

$$\lambda|\xi|^2 \le a_{ij}(x)\xi_i\xi_j \le \Lambda|\xi|^2 \quad \text{for any } x \in B_1 \text{ and } \xi \in \mathbb{R}^n$$

for some positive constants λ and Λ. We consider the function $u \in C^1(B_1)$ satisfying

$$\int_{B_1} a_{ij}D_iuD_j\varphi = 0 \quad \text{for any } \varphi \in C_0^1(B_1).$$

It is easy to check by integration by parts that the harmonic functions satisfy the above equation for $a_{ij} = \delta_{ij}$.

LEMMA 1.36 (Cacciopolli's Inequality) *Suppose $u \in C^1(B_1)$ satisfies*

$$\int_{B_1} a_{ij} D_i u D_j \varphi = 0 \quad \text{for any } \varphi \in C_0^1(B_1).$$

Then for any function $\eta \in C_0^1(B_1)$, we have

$$\int_{B_1} \eta^2 |Du|^2 \leq C \int_{B_1} |D\eta|^2 u^2$$

where C is a positive constant depending only on λ and Λ.

PROOF: For any $\eta \in C_0^1(B_1)$ set $\varphi = \eta^2 u$. Then we have

$$\lambda \int_{B_1} \eta^2 |Du|^2 \leq \Lambda \int_{B_1} \eta |u| |D\eta| |Du|.$$

We obtain the result by the Hölder inequality. □

COROLLARY 1.37 *Let u be as in Lemma 1.36. Then for any $0 \leq r < R \leq 1$ there holds*

$$\int_{B_r} |Du|^2 \leq \frac{C}{(R-r)^2} \int_{B_R} u^2$$

where C is a positive constant depending only on λ and Λ.

PROOF: Take η such that $\eta = 1$ on B_r, $\eta = 0$ outside B_R, and $|D\eta| \leq 2(R-r)^{-1}$. □

COROLLARY 1.38 *Let u be as in Lemma 1.36. Then for any $0 < R \leq 1$ there hold*

$$\int_{B_{R/2}} u^2 \leq \theta \int_{B_R} u^2 \quad \text{and} \quad \int_{B_{R/2}} |Du|^2 \leq \theta \int_{B_R} |Du|^2$$

where $\theta = \theta(n, \lambda, \Lambda) \in (0,1)$.

PROOF: Take $\eta \in C_0^1(B_R)$ with $\eta = 1$ on $B_{R/2}$ and $|D\eta| \leq 2R^{-1}$. Then Lemma 1.36 yields

$$\int_{B_R} |D(\eta u)|^2 \leq C \int_{B_R} |D\eta|^2 u^2 \leq \frac{C}{R^2} \int_{B_R \setminus B_{R/2}} u^2$$

by noting $D\eta = 0$ in $B_{R/2}$. Hence by the Poincaré inequality we get

$$\int_{B_R} (\eta u)^2 \leq c(n) R^2 \int_{B_R} |D(\eta u)|^2.$$

Therefore we obtain

$$\int_{B_{R/2}} u^2 \le C \int_{B_R \setminus B_{R/2}} u^2, \quad \text{which implies} \quad (C+1)\int_{B_{R/2}} u^2 \le C \int_{B_R} u^2.$$

For the second inequality, observe that Lemma 1.36 holds for $u-a$ for arbitrary constant a. Then as before we have

$$\int_{B_R} \eta^2 |Du|^2 \le C \int_{B_R} |D\eta|^2 (u-a)^2 \le \frac{C}{R^2} \int_{B_R \setminus B_{R/2}} (u-a)^2.$$

The Poincaré inequality implies with $a = |B_R \setminus B_{R/2}|^{-1} \int_{B_R \setminus B_{R/2}} u$

$$\int_{B_R \setminus B_{R/2}} (u-a)^2 \le c(n) R^2 \int_{B_R \setminus B_{R/2}} |Du|^2.$$

Hence we obtain

$$\int_{B_{R/2}} |Du|^2 \le C \int_{B_R \setminus B_{R/2}} |Du|^2;$$

in particular,

$$(C+1) \int_{B_{R/2}} |Du|^2 \le C \int_{B_R} |Du|^2.$$

□

REMARK 1.39. Corollary 1.38 implies, in particular, that a harmonic function in $\mathbb{R}^n$ with finite L^2-norm is identically 0 and that a harmonic function in $\mathbb{R}^n$ with finite Dirichlet integral is constant.

REMARK 1.40. By iterating the result in Corollary 1.38, we have the following estimates. Let u be in Lemma 1.36. Then for any $0 < \rho < r \le 1$ there hold

$$\int_{B_\rho} u^2 \le C \left(\frac{\rho}{r}\right)^\mu \int_{B_r} u^2 \quad \text{and} \quad \int_{B_\rho} |Du|^2 \le C \left(\frac{\rho}{r}\right)^\mu \int_{B_r} |Du|^2$$

for some positive constant $\mu = \mu(n, \lambda, \Lambda)$. Later on we will prove that we can take $\mu \in (n-2, n)$ in the second inequality. For harmonic functions we have better results.

LEMMA 1.41 *Suppose $\{a_{ij}\}$ is a constant positive definite matrix with*

$$\lambda |\xi|^2 \le a_{ij} \xi_i \xi_j \le \Lambda |\xi|^2 \quad \textit{for any } \xi \in \mathbb{R}^n$$

for some constants $0 < \lambda \le \Lambda$. Suppose $u \in C^1(B_1)$ satisfies

$$\int_{B_1} a_{ij} D_i u D_j \varphi = 0 \quad \textit{for any } \varphi \in C_0^1(B_1). \tag{1.8}$$

Then for any $0 < \rho \le r$, *there hold*

$$\int_{B_\rho} |u|^2 \le c\left(\frac{\rho}{r}\right)^n \int_{B_r} |u|^2, \tag{1.9}$$

$$\int_{B_\rho} |u - u_\rho|^2 \le c\left(\frac{\rho}{r}\right)^{n+2} \int_{B_r} |u - u_r|^2, \tag{1.10}$$

where $c = c(\lambda, \Lambda)$ *is a positive constant and* u_r *denotes the average of* u *in* B_r.

PROOF: By dilation, consider $r = 1$. We restrict our consideration to the range $\rho \in (0, \frac{1}{2}]$, since (1.9) and (1.10) are trivial for $\rho \in (\frac{1}{2}, 1]$. □

CLAIM.

$$|u|^2_{L^\infty(B_{1/2})} + |Du|^2_{L^\infty(B_{1/2})} \le c(\lambda, \Lambda) \int_{B_1} |u|^2.$$

Therefore for $\rho \in (0, \frac{1}{2}]$

$$\int_{B_\rho} |u|^2 \le \rho^n |u|^2_{L^\infty(B_{1/2})} \le c\rho^n \int_{B_1} |u|^2$$

and

$$\int_{B_\rho} |u - u_\rho|^2 \le \int_{B_\rho} |u - u(0)|^2 \le \rho^{n+2} |Du|^2_{L^\infty(B_{1/2})} \le c\rho^{n+2} \int_{B_1} |u|^2.$$

If u is a solution of (1.8), so is $u - u_1$. With u replaced by $u - u_1$ in the above inequality, there holds

$$\int_{B_\rho} |u - u_\rho|^2 \le c\rho^{n+2} \int_{B_1} |u - u_1|^2.$$

PROOF: We present two methods.

METHOD 1. By rotation, we may assume $\{a_{ij}\}$ is a diagonal matrix. Hence (1.8) becomes

$$\sum_{i=1}^{n} \lambda_i D_{ii} u = 0$$

with $0 < \lambda \le \lambda_i \le \Lambda$ for $i = 1, \dots, n$. It is easy to see there exists an $r_0 = r_0(\lambda, \Lambda) \in (0, \frac{1}{2})$ such that for any $x_0 \in B_{1/2}$ the rectangle

$$\left\{x : \frac{|x_i - x_{0i}|}{\sqrt{\lambda_i}} < r_0\right\}$$

is contained in B_1. Change the coordinate

$$x_i \longmapsto y_i = \frac{x_i}{\sqrt{\lambda_i}} \quad \text{and set} \quad v(y) = u(x).$$

Then v is harmonic in $\{y : \sum_{i=1}^n \lambda_i y_i^2 < 1\}$. In the ball $\{y : |y - y_0| < r_0\}$ use the interior estimates to yield

$$|v(y_0)|^2 + |Dv(y_0)|^2 \leq c(\lambda, \Lambda) \int_{B_{r_0}(y_0)} v^2 \leq c(\lambda, \Lambda) \int_{\{\sum_{i=1}^n \lambda_i y_i^2 < 1\}} v^2.$$

Transform back to u to get

$$|u(x_0)|^2 + |Du(x_0)|^2 \leq c(\lambda, \Lambda) \int_{|x|<1} u^2.$$

METHOD 2. If u is a solution to (1.8), so are any derivatives of u. By applying Corollary 1.37 to derivatives of u we conclude that for any positive integer k

$$\|u\|_{H^k(B_{1/2})} \leq c(k, \lambda, \Lambda)\|u\|_{L^2(B_1)}.$$

If we fix a value of k sufficiently large with respect to n, $H^k(B_{1/2})$ is continuously embedded into $C^1(\bar{B}_{1/2})$ and therefore

$$|u|_{L^\infty(B_{1/2})} + |Du|_{L^\infty(B_{1/2})} \leq c(\lambda, \Lambda)\|u\|_{L^2(B_1)}.$$

This finishes the proof. □

CHAPTER 2

Maximum Principles

2.1. Guide

In this chapter we discuss maximum principles and their applications. Two kinds of maximum principles are discussed. One is due to Hopf and the other to Alexandroff. The former gives the estimates of solutions in terms of the L^∞-norm of the nonhomogeneous terms, while the latter gives the estimates in terms of the L^n-norm. Applications include various a priori estimates and the moving plane method.

Most of the statements in Section 2.2 are rather simple. One probably needs to go over Theorem 2.11 and Proposition 2.13. Section 2.3 is often the starting point of the a priori estimates. Section 2.4 can be omitted in the first reading, as we will look at it again in Section 5.2. The moving plane method explained in Section 2.6 has many recent applications. We choose a very simple example to illustrate such a method. The result goes back to Gidas-Ni-Nirenberg, but the proof contains some recent observations in the paper **[1]**. The classical paper of Gilbarg-Serrin **[7]** may be a very good supplement to this chapter. It may also be a good idea to assume the Harnack inequality of Krylov-Safanov in Section 5.3 and to ask students to improve some of the results in the paper **[7]**.

2.2. Strong Maximum Principle

Suppose Ω is a bounded and connected domain in $\mathbb{R}^n$. Consider the operator L in Ω,

$$Lu \equiv a_{ij}(x)D_{ij}u + b_i(x)D_iu + c(x)u$$

for $u \in C^2(\Omega) \cap C(\bar{\Omega})$. We always assume that a_{ij}, b_i, and c are continuous and hence bounded in $\bar{\Omega}$ and that L is uniformly elliptic in Ω in the following sense:

$$a_{ij}(x)\xi_i\xi_j \geq \lambda|\xi|^2 \quad \text{for any } x \in \Omega \text{ and any } \xi \in \mathbb{R}^n$$

for some positive constant λ.

LEMMA 2.1 *Suppose $u \in C^2(\Omega) \cap C(\bar{\Omega})$ satisfies $Lu > 0$ in Ω with $c(x) \leq 0$ in Ω. If u has a nonnegative maximum in $\bar{\Omega}$, then u cannot attain this maximum in Ω.*

PROOF: Suppose u attains its nonnegative maximum of $\bar{\Omega}$ in $x_0 \in \Omega$. Then $D_iu(x_0) = 0$ and the matrix $B = (D_{ij}(x_0))$ is seminegative definite. By the ellipticity condition the matrix $A = (a_{ij}(x_0))$ is positive definite. Hence the matrix AB is seminegative definite with a nonpositive trace, that is, $a_{ij}(x_0)D_{ij}u(x_0) \leq 0$. This implies $Lu(x_0) \leq 0$, which is a contradiction. □

REMARK 2.2. If $c(x) \equiv 0$, then the requirement for nonnegativeness can be removed. This remark also holds for some results in the rest of this section.

THEOREM 2.3 (Weak Maximum Principle) *Suppose $u \in C^2(\Omega) \cap C(\bar{\Omega})$ satisfies $Lu \geq 0$ in Ω with $c(x) \leq 0$ in Ω. Then u attains on $\partial\Omega$ its nonnegative maximum in $\bar{\Omega}$.*

PROOF: For any $\varepsilon > 0$, consider $w(x) = u(x) + \varepsilon e^{\alpha x_1}$ with α to be determined. Then we have

$$Lw = Lu + \varepsilon e^{\alpha x_1}(a_{11}\alpha^2 + b_1\alpha + c).$$

Since b_1 and c are bounded and $a_{11}(x) \geq \lambda > 0$ for any $x \in \Omega$, by choosing $\alpha > 0$ large enough we get

$$a_{11}(x)\alpha^2 + b_1(x)\alpha + c(x) > 0 \quad \text{for any } x \in \Omega.$$

This implies $Lw > 0$ in Ω. By Lemma 2.1, w attains its nonnegative maximum only on $\partial\Omega$, that is,

$$\sup_{\Omega} w \leq \sup_{\partial\Omega} w^+.$$

Then we obtain

$$\sup_{\Omega} u \leq \sup_{\Omega} w \leq \sup_{\partial\Omega} w^+ \leq \sup \partial\Omega u^+ + \varepsilon \sup_{x \in \partial\Omega} e^{\alpha x_1}.$$

We finish the proof by letting $\varepsilon \to 0$. □

As an application we have the uniqueness of solution $u \in C^2(\Omega) \cap C(\bar{\Omega})$ to the following Dirichlet boundary value problem for $f \in C(\Omega)$ and $\varphi \in C(\partial\Omega)$

$$\begin{aligned} Lu &= f \quad \text{in } \Omega, \\ u &= \varphi \quad \text{on } \partial\Omega, \end{aligned}$$

if $c(x) \leq 0$ in Ω.

REMARK 2.4. The boundedness of domain Ω is essential, since it guarantees the existence of a maximum and a minimum of u in $\bar{\Omega}$. The uniqueness does not hold if the domain is unbounded. Some examples are given in Remark 1.9. Equally important is the nonpositiveness of the coefficient c.

EXAMPLE. Set $\Omega = \{(x, y) \in \mathbb{R}^2 : 0 < x < \pi,\ 0 < y < \pi\}$. Then $u = \sin x \sin y$ is a nontrivial solution for the problem

$$\begin{aligned} \Delta u + 2u &= 0 \quad \text{in } \Omega, \\ u &= 0 \quad \text{on } \partial\Omega. \end{aligned}$$

THEOREM 2.5 (Hopf Lemma) *Let B be an open ball in $\mathbb{R}^n$ with $x_0 \in \partial B$. Suppose $u \in C^2(B) \cap C(B \cup \{x_0\})$ satisfies $Lu \geq 0$ in B with $c(x) \leq 0$ in B. Assume in addition that*

$$u(x) < u(x_0) \quad \textit{for any } x \in B \textit{ and } u(x_0) \geq 0.$$

Then for each outward direction $\boldsymbol{\nu}$ *at* x_0 *with* $\boldsymbol{\nu}\cdot\mathbf{n}(x_0)>0$ *there holds*

$$\liminf_{t\to 0+}\frac{1}{t}[u(x_0)-u(x_0-t\boldsymbol{\nu})]>0.$$

REMARK 2.6. If in addition $u\in C^1(B\cup\{x_0\})$, then we have

$$\frac{\partial u}{\partial \nu}(x_0)>0.$$

PROOF: We may assume that the center of B is at the origin with radius r. We assume further that $u\in C(\bar{B})$ and $u(x)<u(x_0)$ for any $x\in\bar{B}\setminus\{x_0\}$ (since we can construct a tangent ball B_1 to B at x_0 and $B_1\subset B$).

Consider $v(x)=u(x)+\varepsilon h(x)$ for some nonnegative function h. We will choose $\varepsilon>0$ appropriately such that v attains its nonnegative maximum only at x_0. Denote $\Sigma=B\cap B_{1/2r}(x_0)$. Define $h(x)=e^{-\alpha|x|^2}-e^{-\alpha r^2}$ with α to be determined. We check in the following that

$$Lh>0\quad\text{in }\Sigma.$$

Direct calculation yields

$$\begin{aligned}Lh&=e^{-\alpha|x|^2}\left\{4\alpha^2\sum_{i,j=1}^n a_{ij}(x)x_ix_j-2\alpha\sum_{i=1}^n a_{ii}(x)-2\alpha\sum_{n=1}^n b_i(x)x_i+c\right\}-ce^{-\alpha r^2}\\&\geq e^{-\alpha|x|^2}\left\{4\alpha^2\sum_{i,j=1}^n a_{ij}(x)x_ix_j-2\alpha\sum_{i=1}^n[a_{ii}(x)+b_i(x)x_i]+c\right\}.\end{aligned}$$

By the ellipticity assumption, we have

$$\sum_{i,j=1}^n a_{ij}(x)x_ix_j\geq\lambda|x|^2\geq\lambda\left(\frac{r}{2}\right)^2>0\quad\text{in }\Sigma.$$

So for α large enough, we conclude $Lh>0$ in Σ. With such h, we have $Lv=Lu+\varepsilon Lh>0$ in Σ for any $\varepsilon>0$. By Lemma 2.1, v cannot attain its nonnegative maximum inside Σ.

Next we prove that for some small $\varepsilon>0$ v attains at x_0 its nonnegative maximum. Consider v on the boundary $\partial\Sigma$.

- For $x\in\partial\Sigma\cap B$, since $u(x)<u(x_0)$, we have $u(x)<u(x_0)-\delta$ for some $\delta>0$. Take ε small such that $\varepsilon h<\delta$ on $\partial\Sigma\cap B$. Hence for such ε we have $v(x)<u(x_0)$ for $x\in\partial\Sigma\cap B$.
- On $\Sigma\cap\partial B$, $h(x)=0$ and $u(x)<u(x_0)$ for $x\neq x_0$. Hence $v(x)<u(x_0)$ on $\Sigma\cap\partial B\setminus\{x_0\}$ and $v(x_0)=u(x_0)$.

Therefore we conclude

$$\frac{v(x_0)-v(x_0-t\nu)}{t}\geq 0\quad\text{for any small }t>0.$$

Hence we obtain by letting $t\to 0$

$$\liminf_{t\to 0}\frac{1}{t}[u(x_0)-u(x_0-t\nu)]\geq-\varepsilon\frac{\partial h}{\partial\nu}(x_0).$$

By definition of h, we have

$$\frac{\partial h}{\partial \nu}(x_0) < 0.$$

This finishes the proof. □

THEOREM 2.7 (Strong Maximum Principle) *Let $u \in C^2(\Omega) \cap C(\bar{\Omega})$ satisfy $Lu \geq 0$ with $c(x) \leq 0$ in Ω. Then the nonnegative maximum of u in $\bar{\Omega}$ can be assumed only on $\partial\Omega$ unless u is a constant.*

PROOF: Let M be the nonnegative maximum of u in $\bar{\Omega}$. Set $\Sigma = \{x \in \Omega : u(x) = M\}$. It is relatively closed in Ω. We need to show $\Sigma = \Omega$.

We prove by contradiction. If Σ is a proper subset of Ω, then we may find an open ball $B \subset \Omega \setminus \Sigma$ with a point on its boundary belonging to Σ. (In fact, we may choose a point $p \in \Omega \setminus \Sigma$ such that $d(p, \Sigma) < d(p, \partial\Omega)$ first and then extend the ball centered at p. It hits Σ before hitting $\partial\Omega$.) Suppose $x_0 \in \partial B \cap \Sigma$. Obviously we have $Lu \geq 0$ in B and

$$u(x) < u(x_0) \text{ for any } x \in B \quad \text{and} \quad u(x_0) = M \geq 0.$$

Theorem 2.5 implies $\frac{\partial u}{\partial n}(x_0) > 0$ where $\mathbf{n}$ is the outward normal direction at x_0 to the ball B. Since x_0 is the interior maximal point of Ω, $Du(x_0) = 0$. This leads to a contradiction. □

COROLLARY 2.8 (Comparison Principle) *Suppose $u \in C^2(\Omega) \cap C(\bar{\Omega})$ satisfies $Lu \geq 0$ in Ω with $c(x) \leq 0$ in Ω. If $u \leq 0$ on $\partial\Omega$, then $u \leq 0$ in Ω. In fact, either $u < 0$ in Ω or $u \equiv 0$ in Ω.*

In order to discuss the boundary value problem with general boundary condition, we need the following result, which is just a corollary of Theorems 2.5 and 2.7.

COROLLARY 2.9 *Suppose Ω has the interior sphere property and that $u \in C^2(\Omega) \cap C^1(\bar{\Omega})$ satisfies $Lu \geq 0$ in Ω with $c(x) \leq 0$. Assume u attains its nonnegative maximum at $x_0 \in \bar{\Omega}$. Then $x_0 \in \partial\Omega$ and for any outward direction $\boldsymbol{\nu}$ at x_0 to $\partial\Omega$*

$$\frac{\partial u}{\partial \nu}(x_0) > 0$$

unless u is constant in $\bar{\Omega}$.

APPLICATION. Suppose Ω is bounded in $\mathbb{R}^n$ and satisfies the interior sphere property. Consider the the following boundary value problem

$$(*) \qquad \begin{aligned} Lu &= f && \text{in } \Omega \\ \frac{\partial u}{\partial n} + \alpha(x)u &= \varphi && \text{on } \partial\Omega \end{aligned}$$

for some $f \in C(\bar{\Omega})$ and $\varphi \in C(\partial\Omega)$. Assume in addition that $c(x) \leq 0$ in Ω and $\alpha(x) \geq 0$ on $\partial\Omega$. Then problem $(*)$ has a unique solution $u \in C^2(\Omega) \cap C^1(\bar{\Omega})$ if $c \not\equiv 0$ or $\alpha \not\equiv 0$. If $c \equiv 0$ and $\alpha \equiv 0$, problem $(*)$ has a unique solution $u \in C^2(\Omega) \cap C^1(\bar{\Omega})$ up to additive constants.

PROOF: Suppose u is a solution to the following homogeneous equation:

$$Lu = 0 \quad \text{in } \Omega,$$
$$\frac{\partial u}{\partial n} + \alpha(x)u = 0 \quad \text{on } \partial\Omega.$$

CASE 1. $c \not\equiv 0$ or $\alpha \not\equiv 0$. We want to show $u \equiv 0$.

Suppose that u has a positive maximum at $x_0 \in \bar{\Omega}$. If $u \equiv \text{const} > 0$, this contradicts the condition $c \not\equiv 0$ in Ω or $\alpha \not\equiv 0$ on $\partial\Omega$. Otherwise $x_0 \in \partial\Omega$ and $\frac{\partial u}{\partial n}(x_0) > 0$ by Corollary 2.9, which contradicts the boundary value. Therefore $u \equiv 0$.

CASE 2. $c \equiv 0$ and $\alpha \equiv 0$. We want to show $u \equiv \text{const}$.

Suppose u is a nonconstant solution. Then its maximum in $\bar{\Omega}$ is assumed only on $\partial\Omega$ by Theorem 2.7, say at $x_0 \in \partial\Omega$. Again Corollary 2.9 implies $\frac{\partial u}{\partial n}(x_0) > 0$. This is a contradiction. □

The following theorem, due to Serrin, generalizes the comparison principle with no restriction on $c(x)$.

THEOREM 2.10 *Suppose $u \in C^2(\Omega) \cap C(\bar{\Omega})$ satisfies $Lu \geq 0$. If $u \leq 0$ in Ω, then either $u < 0$ in Ω or $u \equiv 0$ in Ω.*

PROOF: We present two methods.

METHOD 1. Suppose $u(x_0) = 0$ for some $x_0 \in \Omega$. We will prove that $u \equiv 0$ in Ω.

Write $c(x) = c^+(x) - c^-(x)$ where $c^+(x)$ and $c^-(x)$ are the positive part and negative part of $c(x)$, respectively. Hence u satisfies

$$a_{ij}D_{ij}u + b_i D_i u - c^- u \geq -c^+ u \geq 0.$$

So we have $u \equiv 0$ by Theorem 2.7.

METHOD 2. Set $v = ue^{-\alpha x_1}$ for some $\alpha > 0$ to be determined. By $Lu \geq 0$, we have

$$a_{ij}D_{ij}v + [\alpha(a_{1i} + a_{i1}) + b_i]D_i v + (a_{11}\alpha^2 + b_1\alpha + c)v \geq 0.$$

Choose α large enough such that $a_{11}\alpha^2 + b_1\alpha + c > 0$. Therefore v satisfies

$$a_{ij}D_{ij}v + [\alpha(a_{1i} + a_{i1}) + b_i]D_i v \geq 0.$$

Hence we apply Theorem 2.7 to v to conclude that either $v < 0$ in Ω or $v \equiv 0$ in Ω. □

The next result is the general maximum principle for the operator L with no restriction on $c(x)$.

THEOREM 2.11 *Suppose there exists a $w \in C^2(\Omega) \cap C^1(\bar{\Omega})$ satisfying $w > 0$ in $\bar{\Omega}$ and $Lw \leq 0$ in Ω. If $u \in C^2(\Omega) \cap C(\bar{\Omega})$ satisfies $Lu \geq 0$ in Ω, then $\frac{u}{w}$ cannot assume in Ω its nonnegative maximum unless $\frac{u}{w} \equiv$ const. If, in addition,*

$\frac{u}{w}$ *assumes its nonnegative maximum at* $x_0 \in \partial\Omega$ *and* $\frac{u}{w} \not\equiv$ const, *then for any outward direction* $\mathbf{v}$ *at* x_0 *to* $\partial\Omega$ *there holds*

$$\frac{\partial}{\partial\nu}\left(\frac{u}{w}\right)(x_0) > 0$$

if $\partial\Omega$ *has the interior sphere property at* x_0.

PROOF: Set $v = \frac{u}{w}$. Then v satisfies

$$a_{ij}D_{ij}v + B_iD_iv + \left(\frac{Lw}{w}\right)v \geq 0$$

where $B_i = b_i + \frac{2}{w}a_{ij}D_{ij}w$. We may apply Theorem 2.7 and Corollary 2.9 to v. □

REMARK 2.12. If the operator L in Ω satisfies the condition of Theorem 2.11, then the comparison principle applies to L. In particular, the Dirichlet boundary value problem

$$\begin{aligned} Lu &= f \quad \text{in } \Omega, \\ u &= \varphi \quad \text{on } \partial\Omega, \end{aligned}$$

has at most one solution.

The next result is the so-called maximum principle for a *narrow domain.*

PROPOSITION 2.13 *Let* d *be a positive number and* $\mathbf{e}$ *be a unit vector such that* $|(y-x)\cdot\mathbf{e}| < d$ *for any* $x, y \in \Omega$. *Then there exists a* $d_0 > 0$, *depending only on* λ *and the sup-norm of* b_i *and* c^+, *such that the assumptions of Theorem* 2.11 *are satisfied if* $d \leq d_0$.

PROOF: By choosing $\mathbf{e} = (1, 0, \ldots, 0)$ we suppose $\bar{\Omega}$ lies in $\{0 < x_1 < d\}$. Assume in addition $|b_i|, c^+ \leq N$ for some positive constant N. We construct w as follows. Set $w = e^{\alpha d} - e^{\alpha x_1} > 0$ in $\bar{\Omega}$. By direct calculation we have

$$Lw = -(a_{11}\alpha^2 + b_1\alpha)e^{\alpha x_1} + c(e^{\alpha d} - e^{\alpha x_1}) \leq -(a_{11}\alpha^2 + b_1\alpha) + Ne^{\alpha d}.$$

Choose α so large that

$$a_{11}\alpha^2 + b_1\alpha \geq \lambda\alpha^2 - N\alpha \geq 2N.$$

Hence $Lw \leq -2N + Ne^{\alpha d} = N(e^{\alpha d} - 2) \leq 0$ if d is small such that $e^{\alpha d} \leq 2$. □

REMARK 2.14. Some results in this section, including Proposition 2.13, hold for unbounded domain. Compare Proposition 2.13 with Theorem 2.32.

2.3. A Priori Estimates

In this section we derive a priori estimates for solutions to the Dirichlet problem and the Neumann problem.

Suppose Ω is a bounded and connected domain in $\mathbb{R}^n$. Consider the operator L in Ω

$$Lu \equiv a_{ij}(x)D_{ij}u + b_i(x)D_iu + c(x)u$$

for $u \in C^2(\Omega) \cap C(\bar{\Omega})$. We assume that a_{ij}, b_i, and c are continuous and hence bounded in $\bar{\Omega}$ and that L is uniformly elliptic in Ω, that is,

$$a_{ij}(x)\xi_i\xi_j \geq \lambda|\xi|^2 \quad \text{for any } x \in \Omega \text{ and any } \xi \in \mathbb{R}^n$$

where λ is a positive number. We denote by Λ the sup-norm of a_{ij} and b_i, that is,

$$\max_{\Omega} |a_{ij}| + \max_{\Omega} |b_i| \leq \Lambda.$$

PROPOSITION 2.15 *Suppose $u \in C^2(\Omega) \cap C(\bar{\Omega})$ satisfies*

$$\begin{cases} Lu = f & \text{in } \Omega, \\ u = \varphi & \text{on } \partial\Omega, \end{cases}$$

for some $f \in C(\bar{\Omega})$ and $\varphi \in C(\partial\Omega)$. If $c(x) \leq 0$, then there holds

$$|u(x)| \leq \max_{\partial\Omega} |\varphi| + C \max_{\Omega} |f| \quad \text{for any } x \in \Omega$$

where C is a positive constant depending only on λ, Λ, and $\operatorname{diam}(\Omega)$.

PROOF: We will construct a function w in Ω such that

$$\text{(i)} \qquad L(w \pm u) = Lw \pm f \leq 0 \quad \text{or} \quad Lw \leq \mp f \quad \text{in } \Omega,$$

$$\text{(ii)} \qquad w \pm u = w \pm \varphi \geq 0 \quad \text{or} \quad w \geq \mp\varphi \quad \text{on } \partial\Omega.$$

Denote $F = \max_\Omega |f|$ and $\Phi = \max_{\partial\Omega} |\varphi|$. We need

$$\begin{aligned} Lw &\leq -F \quad \text{in } \Omega, \\ w &\geq \Phi \quad \text{on } \partial\Omega. \end{aligned}$$

Suppose the domain Ω lies in the set $\{0 < x_1 < d\}$ for some $d > 0$. Set $w = \Phi + (e^{\alpha d} - e^{\alpha x_1})F$ with $\alpha > 0$ to be chosen later. Then we have by direct calculation

$$\begin{aligned} -Lw &= (a_{11}\alpha^2 + b_1\alpha)Fe^{\alpha x_1} - c\Phi - c(e^{\alpha d} - e^{\alpha x_1})F \\ &\geq (a_{11}\alpha^2 + b_1\alpha)F \geq (\alpha^2\lambda + b_1\alpha)F \geq F \end{aligned}$$

by choosing α large such that $\alpha^2\lambda + b_1(x)\alpha \geq 1$ for any $x \in \Omega$. Hence w satisfies (i) and (ii). By Corollary 2.8 (the comparison principle) we conclude $-w \leq u \leq w$ in Ω; in particular,

$$\sup_{\Omega} |u| \leq \Phi + (e^{\alpha d} - 1)F$$

where α is a positive constant depending only on λ and Λ. □

PROPOSITION 2.16 *Suppose $u \in C^2(\Omega) \cap C^1(\bar{\Omega})$ satisfies*

$$\begin{cases} Lu = f & \text{in } \Omega, \\ \frac{\partial u}{\partial n} + \alpha(x)u = \varphi & \text{on } \partial\Omega, \end{cases}$$

where $\mathbf{n}$ is the outward normal direction to $\partial\Omega$. If $c(x) \leq 0$ in Ω and $\alpha(x) \geq \alpha_0 > 0$ on $\partial\Omega$, then there holds

$$|u(x)| \leq C\{\max_{\partial\Omega} |\varphi| + \max_{\Omega} |f|\} \quad \text{for any } x \in \Omega$$

where C is a positive constant depending only on λ, Λ, α_0, and diam(Ω).

PROOF: We prove for a special case and the general case.

CASE 1. Special case: $c(x) \le -c_0 < 0$.

We will show

$$|u(x)| \le \frac{1}{c_0} F + \frac{1}{\alpha_0} \Phi \quad \text{for any } x \in \Omega.$$

Define $v = \frac{1}{c_0} F + \frac{1}{\alpha_0} \Phi \pm u$. Then we have

$$Lv = c(x)\left(\frac{1}{c_0} F + \frac{1}{\alpha_0} \Phi\right) \pm f \le -F \pm f \le 0 \quad \text{in } \Omega,$$

$$\frac{\partial v}{\partial n} + \alpha v = \alpha\left(\frac{1}{c_0} F + \frac{1}{\alpha_0}\Phi\right) \pm \varphi \ge \Phi \pm \varphi \ge 0 \qquad \text{on } \partial\Omega.$$

If v has a negative minimum in $\bar{\Omega}$, then v attains it on $\partial\Omega$ by Theorem 2.5, say, at $x_0 \in \partial\Omega$. This implies $\frac{\partial v}{\partial n}(x_0) \le 0$ for $\mathbf{n} = \mathbf{n}(x_0)$, the outward normal direction at x_0. Therefore we get

$$\left(\frac{\partial v}{\partial n} + \alpha v\right)(x_0) \le \alpha v(x_0) < 0,$$

which is a contradiction. Hence we have $v \ge 0$ in $\bar{\Omega}$, in particular,

$$|u(x)| \le \frac{1}{c_0} F + \frac{1}{\alpha_0} \Phi \quad \text{for any } x \in \Omega.$$

Note that for this special case c_0 and α_0 are independent of λ and Λ.

CASE 2. General case: $c(x) \le 0$ for any $x \in \Omega$.

Consider the auxiliary function $u(x) = z(x)w(x)$ where z is a positive function in $\bar{\Omega}$ to be determined. Direct calculation shows that w satisfies

$$a_{ij} D_{ij} w + B_i D_i w + \left(c + \frac{a_{ij} D_{ij} z + b_i D_i z}{z}\right) w = \frac{f}{z} \quad \text{in } \Omega,$$

$$\frac{\partial w}{\partial n} + \left(\alpha + \frac{1}{z}\frac{\partial z}{\partial n}\right) w = \frac{\varphi}{z} \quad \text{on } \partial\Omega,$$

where $B_i = \frac{1}{z}(a_{ij} + a_{ji}) D_j z + b_i$. We need to choose the function $z > 0$ in $\bar{\Omega}$ such that there hold in

$$c + \frac{a_{ij} D_{ij} z + b_i D_i z}{z} \le -c_0(\lambda, \Lambda, d, \alpha_0) < 0 \quad \text{in } \Omega,$$

$$\alpha + \frac{1}{z}\frac{\partial z}{\partial n} \ge \frac{1}{2}\alpha_0 \qquad \text{on } \partial\Omega,$$

or

$$\frac{a_{ij} D_{ij} z + b_i D_i z}{z} \le -c_0 < 0 \quad \text{in } \Omega,$$

$$\left|\frac{1}{z}\frac{\partial z}{\partial n}\right| \le \frac{1}{2}\alpha_0 \qquad \text{on } \partial\Omega.$$

Suppose the domain Ω lies in $\{0 < x_1 < d\}$. Choose $z(x) = A + e^{\beta d} - e^{\beta x_1}$ for $x \in \Omega$ for some positive A and β to be determined. Direct calculation shows

$$-\frac{1}{z}(a_{ij}D_{ij}z + b_iD_iz) = \frac{(\beta^2 a_{11} + \beta b_1)e^{\beta x_1}}{A + e^{\beta d} - e^{\beta x_1}}$$
$$\geq \frac{\beta^2 a_{11} + \beta b_1}{A + e^{\beta d}} \geq \frac{1}{A + e^{\beta d}} > 0$$

if β is chosen such that $\beta^2 a_{11} + \beta b_1 \geq 1$. Then we have

$$\left|\frac{1}{z}\frac{\partial z}{\partial n}\right| \leq \frac{\beta}{A}e^{\beta d} \leq \frac{1}{2}\alpha_0$$

if A is chosen large. This reduces to the special case we just discussed. The new extra first-order term does not change the result. We may apply the special case to w. □

REMARK 2.17. The result fails if we just assume $\alpha(x) \geq 0$ on $\partial\Omega$. In fact, we cannot even get the uniqueness.

2.4. Gradient Estimates

The basic idea in the treatment of gradient estimates, due to Bernstein, involves differentiation of the equation with respect to x_k, $k = 1, \dots, n$, followed by multiplication by $D_k u$ and summation over k. The maximum principle is then applied to the resulting equation in the function $v = |Du|^2$, possibly with some modification. There are two kinds of gradient estimates, global gradient estimates and interior gradient estimates. We will use semilinear equations to illustrate the idea.

Suppose Ω is a bounded and connected domain in $\mathbb{R}^n$. Consider the equation

$$a_{ij}(x)D_{ij}u + b_i(x)D_iu = f(x,u) \quad \text{in } \Omega$$

for $u \in C^2(\Omega)$ and $f \in C(\Omega \times \mathbb{R})$. We always assume that a_{ij} and b_i are continuous and hence bounded in $\bar{\Omega}$ and that the equation is uniformly elliptic in Ω in the following sense:

$$a_{ij}(x)\xi_i\xi_j \geq \lambda|\xi|^2 \quad \text{for any } x \in \Omega \text{ and any } \xi \in \mathbb{R}^n$$

for some positive constant λ.

PROPOSITION 2.18 *Suppose $u \in C^3(\Omega) \cap C^1(\bar{\Omega})$ satisfies*

$$a_{ij}(x)D_{ij}u + b_i(x)D_iu = f(x,u) \quad \textit{in } \Omega \tag{2.1}$$

for $a_{ij}, b_i \in C^1(\bar{\Omega})$ and $f \in C^1(\bar{\Omega} \times \mathbb{R})$. Then there holds

$$\sup_{\Omega}|Du| \leq \sup_{\partial\Omega}|Du| + C$$

where C is a positive constant depending only on λ, diam(Ω), $|a_{ij}, b_i|_{C^1(\bar{\Omega})}$, $M = |u|_{L^\infty(\Omega)}$, *and* $|f|_{C^1(\bar{\Omega}\times[-M,M])}$.

PROOF: Set $L \equiv a_{ij}D_{ij} + b_iD_i$. We calculate $L(|Du|^2)$ first. Note

$$D_i(|Du|^2) = 2D_kuD_{ki}u,$$

$$D_{ij}(|Du|^2) = 2D_{ki}D_{kj}u + 2D_kuD_{kij}u. \tag{2.2}$$

Differentiating (2.1) with respect to x_k, multiplying by D_ku, and summing over k, we have by (2.2)

$$\begin{aligned} a_{ij}D_{ij}(|Du|^2) + b_iD_i(|Du|^2) \\ &= 2a_{ij}D_{ki}uD_{kj}u - 2D_ka_{ij}D_kuD_{ij}u \\ &\quad - 2D_kb_iD_kuD_iu + 2D_zf|Du|^2 + 2D_kfD_ku. \end{aligned}$$

The ellipticity assumption implies

$$\sum_{i,j,k} a_{ij}D_{ki}uD_{kj}u \geq \lambda|D^2u|^2.$$

By the Cauchy inequality, we have

$$L(|Du|^2) \geq \lambda|D^2u|^2 - C|Du|^2 - C$$

with C a positive constant depending only on $|f|_{C^1(\bar{\Omega}\times[-M,M])}$, $|a_{ij}, b_i|_{C^1(\bar{\Omega})}$, and λ.

We need to add another term u^2. We have by the ellipticity assumption

$$\begin{aligned} L(u^2) &= 2a_{ij}D_iuD_ju + 2u\{a_{ij}D_{ij}u + b_iD_iu\} \\ &\geq 2\lambda|Du|^2 + 2uf. \end{aligned}$$

Therefore we obtain

$$\begin{aligned} L(|Du|^2 + \alpha u^2) &\geq \lambda|D^2u|^2 + (2\lambda\alpha - C)|Du|^2 - C \\ &\geq \lambda|D^2u|^2 + |Du|^2 - C \end{aligned}$$

if we choose $\alpha > 0$ large, with C depending in addition on M. In order to control the constant term we may consider another function $e^{\beta x_1}$ for $\beta > 0$. Hence we get

$$\begin{aligned} L(|Du|^2 + \alpha u^2 + e^{\beta x_1}) &\geq \lambda|D^2u|^2 + |Du|^2 \\ &\quad + \{\beta^2 a_{11}e^{\beta x_1} + \beta b_1 e^{\beta x_1} - C\}. \end{aligned}$$

If we put the domain $\Omega \subset \{x_1 > 0\}$, then $e^{\beta x_1} \geq 1$ for any $x \in \Omega$. By choosing β large, we may make the last term positive. Therefore, if we set $w = |Du|^2 + \alpha|u|^2 + e^{\beta x_1}$ for large α, β depending only on λ, $\operatorname{diam}(\Omega)$, $|a_{ij}, b_i|_{C^1(\bar{\Omega})}$, $M = |u|_{L^\infty(\Omega)}$, and $|f|_{C^1(\bar{\Omega}\times[-M,M])}$, then we obtain

$$Lw \geq 0 \quad \text{in } \Omega.$$

By the maximum principle we have

$$\sup_\Omega w \leq \sup_{\partial\Omega} w.$$

This finishes the proof. □

Similarly, we can discuss the interior gradient bound. In this case, we just require the bound of $\sup_\Omega |u|$.

PROPOSITION 2.19 *Suppose $u \in C^3(\Omega)$ satisfies*

$$a_{ij}(x)D_{ij}u + b_i(x)D_i u = f(x,u) \quad in\ \Omega$$

for $a_{ij}, b_i \in C^1(\bar{\Omega})$ and $f \in C^1(\bar{\Omega}\times\mathbb{R})$. Then there holds for any compact subset $\Omega' \Subset \Omega$

$$\sup_{\Omega'} |Du| \le C$$

where C is a positive constant that depends only on λ, diam(Ω), dist$(\Omega', \partial\Omega)$, *$|a_{ij}, b_i|_{C^1(\bar{\Omega})}$, $M = |u|_{L^\infty(\Omega)}$, and $|f|_{C^1(\bar{\Omega}\times[-M,M])}$.*

PROOF: We need to take a cutoff function $\gamma \in C_0^\infty(\Omega)$ with $\gamma \ge 0$ and consider the auxiliary function with the following form:

$$w = \gamma|Du|^2 + \alpha|u|^2 + e^{\beta x_1}.$$

Set $v = \gamma|Du|^2$. Then we have for operator L defined as before

$$Lv = (L\gamma)|Du|^2 + \gamma L(|Du|^2) + 2a_{ij}D_i\gamma D_j|Du|^2.$$

Recall an inequality in the proof of Proposition 2.18,

$$L(|Du|^2) \ge \lambda|D^2u|^2 - C|Du|^2 - C.$$

Hence we have

$$Lv \ge \lambda\gamma|D^2u|^2 + 2a_{ij}D_kuD_i\gamma D_{kj}u - C|Du|^2 + (L\gamma)|Du|^2 - C.$$

The Cauchy inequality implies for any $\varepsilon > 0$

$$|2a_{ij}D_kuD_i\gamma D_{kj}u| \le \varepsilon|D\gamma|^2|D^2u|^2 + c(\varepsilon)|Du|^2.$$

For the cutoff function γ, we require that

$$|D\gamma|^2 \le C\gamma \quad \text{in } \Omega.$$

Therefore we have by taking $\varepsilon > 0$ small

$$\begin{aligned} Lv &\ge \lambda\gamma|D^2u|^2\left(1 - \varepsilon\frac{|D\gamma|^2}{\gamma}\right) - C|Du|^2 - C \\ &\ge \frac{1}{2}\lambda\gamma|D^2u|^2 - C|Du|^2 - C. \end{aligned}$$

Now we may proceed as before. □

In the rest of this section we use barrier functions to derive the boundary gradient estimates. We need to assume that the domain Ω satisfies the uniform exterior sphere property.

PROPOSITION 2.20 *Suppose $u \in C^2(\Omega) \cap C(\bar{\Omega})$ satisfies*

$$a_{ij}(x)D_{ij}u + b_i(x)D_i u = f(x,u) \quad in\ \Omega$$

for $a_{ij}, b_i \in C(\bar{\Omega})$ and $f \in C(\bar{\Omega} \times \mathbb{R})$. Then there holds

$$|u(x) - u(x_0)| \le C|x - x_0| \quad for\ any\ x \in \Omega\ and\ x_0 \in \partial\Omega$$

where C is a positive constant depending only on λ, Ω, $|a_{ij}, b_i|_{L^\infty(\Omega)}$, $M = |u|_{L^\infty(\Omega)}$, $|f|_{L^\infty(\Omega\times[-M,M])}$, and $|\varphi|_{C^2(\bar\Omega)}$ for some $\varphi \in C^2(\bar\Omega)$ with $\varphi = u$ on $\partial\Omega$.

PROOF: For simplicity we assume $u = 0$ on $\partial\Omega$. As before, set $L = a_{ij}D_{ij} + b_i D_i$. Then we have

$$L(\pm u) = \pm f \geq -F \quad \text{in } \Omega$$

where we denote $F = \sup_\Omega |f(\cdot, u)|$. Now fix $x_0 \in \partial\Omega$. We will construct a function w such that

$$Lw \leq -F \ \text{ in } \Omega, \quad w(x_0) = 0, \ \ w|_{\partial\Omega} \geq 0.$$

Then by the maximum principle we have

$$-w \leq u \leq w \quad \text{in } \Omega.$$

Taking the normal derivative at x_0, we have

$$\left|\frac{\partial u}{\partial n}(x_0)\right| \leq \frac{\partial w}{\partial n}(x_0).$$

So we need to bound $\frac{\partial w}{\partial n}(x_0)$ independently of x_0.

Consider the exterior ball $B_R(y)$ with $\bar B_R(y) \cap \bar\Omega = \{x_0\}$. Define $d(x)$ as the distance from x to $\partial B_R(y)$. Then we have

$$0 < d(x) < D \equiv \operatorname{diam}(\Omega) \quad \text{for any } x \in \Omega.$$

In fact, $d(x) = |x - y| - R$ for any $x \in \Omega$. Consider $w = \psi(d)$ for some function defined in $[0, \infty)$. Then we need

$$\begin{aligned} &\psi(0) = 0 && (\Longrightarrow w(x_0) = 0) \\ &\psi(d) > 0 \ \text{ for } d > 0 && (\Longrightarrow w|_{\partial\Omega} \geq 0) \\ &{}'(0) \text{ is controlled.} && \end{aligned}$$

From the first two inequalities, it is natural to require that $\psi'(d) > 0$. Note

$$Lw = {}''a_{ij} D_i d D_j d + {}'a_{ij} D_{ij} d + {}'b_i D_i d.$$

Direct calculation yields

$$D_i d(x) = \frac{x_i - y_i}{|x - y|},$$

$$D_{ij} d(x) = \frac{\delta_{ij}}{|x - y|} - \frac{(x_i - y_i)(x_i - y_i)}{|x - y|^3},$$

which imply $|Dd| = 1$ and with $\Lambda = \sup |a_{ij}|$

$$\begin{aligned} a_{ij} D_{ij} d = \frac{a_{ii}}{|x - y|} - \frac{a_{ij}}{|x - y|} D_i d D_j d &\leq \frac{n\Lambda}{|x - y|} - \frac{\lambda}{|x - y|} \\ &= \frac{n\Lambda - \lambda}{|x - y|} \leq \frac{n\Lambda - \lambda}{R}. \end{aligned}$$

Therefore we obtain by ellipticity

$$\begin{aligned} Lw &\le {}''a_{ij}D_idD_jd + {}'\Big(\frac{n\Lambda-\lambda}{R}+|b|\Big)\\ &\le \lambda\psi'' + \psi'\Big(\frac{n\Lambda-\lambda}{R}+|b|\Big) \end{aligned}$$

if we require $\psi''<0$. Hence in order to have $Lw\le -F$ we need

$$\lambda\psi'' + {}'\Big(\frac{n\Lambda-\lambda}{R}+|b|\Big)+F\le 0.$$

To this end, we study the equation for some positive constants a and b

$${}'' + a\psi' + b = 0$$

whose solution is given by

$$\psi(d) = -\frac{b}{a}d + \frac{C_1}{a} - \frac{C_2}{a}e^{-ad}$$

for some constants C_1 and C_2. For $\psi(0)=0$, we need $C_1=C_2$. Hence we have for some constant C

$$\psi(d) = -\frac{b}{a}\,d + \frac{C}{a}(1-e^{-ad}),$$

which implies

$$\begin{aligned} {}'(d) &= Ce^{-ad} - \frac{b}{a} = e^{-ad}\Big(C-\frac{b}{a}e^{ad}\Big)\\ {}''(d) &= -Cae^{-ad}. \end{aligned}$$

In order to have $\psi'(d)>0$, we need $C\ge \frac{b}{a}e^{aD}$. Since $\psi'(d)>0$ for $d>0$, so $\psi(d)>\psi(0)=0$ for any $d>0$. Therefore we take

$$\begin{aligned} \psi(d) &= -\frac{b}{a}\,d + \frac{b}{a^2}e^{aD}(1-e^{-ad})\\ &= \frac{b}{a}\Big\{\frac{1}{a}e^{aD}(1-e^{-ad}) - d\Big\}. \end{aligned}$$

Such ψ satisfies all the requirements we imposed. This finishes the proof. □

2.5. Alexandroff Maximum Principle

Suppose Ω is a bounded domain in $\mathbb{R}^n$ and consider a second-order elliptic operator L in Ω

$$L \equiv a_{ij}(x)D_{ij} + b_i(x)D_i + c(x)$$

where coefficients a_{ij}, b_i, c are at least continuous in Ω. Ellipticity means that the coefficient matrix $A=(a_{ij})$ is positive definite everywhere in Ω. We set $D=\det(A)$ and $D^*=D^{1/n}$ so that D^* is the geometric mean of the eigenvalues of A. Throughout this section we assume

$$0<\lambda\le D^*\le\Lambda$$

where λ and Λ are two positive constants, which denote, respectively, the minimal and maximal eigenvalues of A.

Before stating the main theorem, we first introduce the concept of contact sets. For $u \in C^2(\Omega)$ we define

$$\Gamma^+ = \{y \in \Omega : u(x) \le u(y) + Du(y) \cdot (x - y) \text{ for any } x \in \Omega\}.$$

The set Γ^+ is called the upper contact set of u, and the Hessian matrix $D^2u = (D_{ij}u)$ is nonpositive on Γ^+. In fact, the upper contact set can also be defined for continuous function u by the following:

$$\Gamma^+ = \{y \in \Omega : u(x) \le u(y) + p \cdot (x - y) \text{ for any } x \in \Omega \\ \text{and some } p = p(y) \in \mathbb{R}^n\}.$$

Clearly, u is concave if and only if $\Gamma^+ = \Omega$. If $u \in C^1(\Omega)$, then $p(y) = Du(y)$, and any support hyperplane must then be a tangent plane to the graph.

Now we consider the equation of the form

$$Lu = f \quad \text{in } \Omega$$

for some $f \in C(\Omega)$.

THEOREM 2.21 *Suppose $u \in C(\bar{\Omega}) \cap C^2(\Omega)$ satisfies $Lu \ge f$ in Ω with the following conditions*:

$$\frac{|\boldsymbol{b}|}{D^*}, \frac{f}{D^*} \in L^n(\Omega) \quad \text{and} \quad c \le 0 \text{ in } \Omega.$$

Then there holds

$$\sup_\Omega u \le \sup_{\partial\Omega} u^+ + C \left\| \frac{f^-}{D^*} \right\|_{L^n(\Gamma^+)}$$

where Γ^+ is the upper contact set of u and C is a constant depending only on n, $\operatorname{diam}(\Omega)$, *and* $\|\frac{\boldsymbol{b}}{D^*}\|_{L^n(\Gamma^+)}$. *In fact, C can be written as*

$$d \cdot \left\{ \exp\left\{ \frac{2^{n-2}}{\omega_n n^n} \left(\left\| \frac{\boldsymbol{b}}{D^*} \right\|^n_{L^n(\Gamma^+)} + 1 \right) \right\} - 1 \right\}$$

with ω_n as the volume of the unit ball in $\mathbb{R}^n$. Here $\boldsymbol{b} = (b_1, b_2, \dots, b_n)$.

REMARK 2.22. The integral domain Γ^+ can be replaced by

$$\Gamma^+ \cap \{x \in \Omega : u(x) > \sup_{\partial\Omega} u^+\}.$$

REMARK 2.23. There is no assumption on uniform ellipticity. Compare with Hopf's maximum principle in Section 1.

We need a lemma first.

LEMMA 2.24 *Suppose $g \in L^1_{\text{loc}}(\mathbb{R}^n)$ is nonnegative. Then for any $u \in C(\bar{\Omega}) \cap C^2(\Omega)$ there holds*

$$\int_{B_{\tilde{M}}(0)} g \le \int_{\Gamma^+} g(Du)|\det D^2u|$$

where Γ^+ *is the upper contact set of* u *and* $\widetilde{M} = (\sup_\Omega u - \sup_{\partial\Omega} u^+)/d$ *with* $d = \operatorname{diam}(\Omega)$.

REMARK 2.25. For any positive definite matrix $A = (a_{ij})$ we have

$$\det(-D^2u) \le \frac{1}{D}\left(\frac{-a_{ij}D_{ij}u}{n}\right)^n \quad \text{on } \Gamma^+.$$

Hence we have another form for Lemma 4.2

$$\int_{B_{\widetilde{M}}(0)} g \le \int_{\Gamma^+} g(Du)\left(\frac{-a_{ij}D_{ij}u}{nD^*}\right)^n.$$

REMARK 2.26. A special case corresponds to $g = 1$:

$$\begin{aligned}\sup_\Omega \sup u &\le \sup_{\partial\Omega} u^+ + \frac{d}{\omega_n^{1/n}}\left(\int_{\Gamma^+} |\det D^2u|\right)^{1/n}\\ &\le \sup_{\partial\Omega} u^+ + \frac{d}{\omega_n^{1/n}}\left(\int_{\Gamma^+}\left(-\frac{a_{ij}D_{ij}u}{nD^*}\right)^n\right)^{1/n}.\end{aligned}$$

Note that this is Theorem 2.21 if $b_i \equiv 0$ and $c \equiv 0$.

PROOF OF LEMMA 2.24: Without loss of generality we assume $u \le 0$ on $\partial\Omega$. Set $\Omega^+ = \{u > 0\}$. By the area formula for Du in $\Gamma^+ \cap \Omega^+ \subset \Omega$, we have

$$\int_{Du(\Gamma^+\cap\Omega^+)} g \le \int_{\Gamma^+\cap\Omega^+} g(Du)|\det(D^2u)|, \tag{2.3}$$

where $|\det(D^2u)|$ is the Jacobian of the map $Du : \Omega \to \mathbb{R}^n$. In fact, we may consider $\chi_\varepsilon = Du - \varepsilon\,\mathrm{Id} : \Omega \to \mathbb{R}^n$. Then $D\chi_\varepsilon = D^2u - \varepsilon I$, which is negative definite in Γ^+. Hence by the change-of-variable formula we have

$$\int_{\chi_\varepsilon(\Gamma^+\cap\Omega^+)} g = \int_{\Gamma^+\cap\Omega^+} g(\chi_\varepsilon)|\det(D^2u - \varepsilon I)|,$$

which implies (2.3) if we let $\varepsilon \to 0$.

Now we claim $B_{\widetilde{M}}(0) \subset Du(\Gamma^+\cap\Omega^+)$, that is, for any $a \in \mathbb{R}^n$ with $|a| < \widetilde{M}$ there exists $x \in \Gamma^+ \cap \Omega^+$ such that $a = Du(x)$.

We may assume u attains its maximum $m > 0$ at $0 \in \Omega$, that is,

$$u(0) = m = \sup \Omega u.$$

Consider an affine function for $|a| < \frac{m}{d}\ (\equiv \widetilde{M})$

$$L(x) = m + a \cdot x.$$

Then $L(x) > 0$ for any $x \in \Omega$ and $L(0) = m$. Since u assumes its maximum at 0, then $Du(0) = 0$. Hence there exists an x_1 close to 0 such that $u(x_1) > L(x_1) > 0$. Note that $u \le 0 < L$ on $\partial\Omega$. Hence there exists an $\widetilde{x} \in \Omega$ such that $Du(\widetilde{x}) = DL(\widetilde{x}) = a$. Now we may translate vertically the plane $y = L(x)$ to the highest such position, that is, the whole surface $y = u(x)$ lies below the plane. Clearly at such a point, the function u is positive. □

PROOF OF THEOREM 2.21: We should choose g appropriately in order to apply Lemma 2.24. Note if $f \equiv 0$ and $c \equiv 0$ then $(-a_{ij}D_{ij}u)^n \leq |b|^n|Du|^n$ in Ω. This suggests that we should take $g(p) = |p|^{-n}$. However, such a function is not locally integrable (at the origin). Hence we will choose $g(p) = (|p|^n + \mu^n)^{-1}$ and then let $\mu \to 0^+$.

First we have by the Cauchy inequality

$$\begin{aligned}
-a_{ij}D_{ij} &\leq b_i D_i u + cu - f \\
&\leq b_i D_i u - f \quad \text{in } \Omega^+ = \{x : u(x) > 0\} \\
&\leq |b| \cdot |Du| + f^- \\
&\leq \left(|b|^n + \frac{(f^-)^n}{\mu^n}\right)^{1/n} \cdot (|Du|^n + \mu^n)^{1/n} \cdot (1+1)^{\frac{n-2}{n}};
\end{aligned}$$

in particular,

$$(-a_{ij}D_{ij}u)^n \leq \left(|b|^n + \left(\frac{f^-}{\mu}\right)^n\right)(|Du|^n + \mu^n) \cdot 2^{n-2}.$$

Now we choose

$$g(p) = \frac{1}{|p|^n + \mu^n}.$$

By Lemma 2.24 we have

$$\int_{B_{\widetilde{M}}(0)} g \leq \frac{2^{n-2}}{n^n} \int_{\Gamma^+ \cap \Omega^+} \frac{|b^n| + \mu^{-n}(f^-)^n}{D}.$$

We evaluate the integral in the left-hand side in the following way:

$$\begin{aligned}
\int_{B_{\widetilde{M}}(0)} g = \omega_n \int_0^{\widetilde{M}} \frac{r^{n-1}}{r^n + \mu^n} dr &= \frac{\omega_n}{n} \log \frac{\widetilde{M}^n + \mu^n}{\mu^n} \\
&= \frac{\omega_n}{n} \log\left(\frac{\widetilde{M}^n}{\mu^n} + 1\right).
\end{aligned}$$

Therefore we obtain

$$\widetilde{M}^n \leq \mu^n \left\{ \exp\left\{ \frac{2^{n-2}}{\omega_n n^n} \left[\left\| \frac{b}{D^*} \right\|^n_{L^n(\Gamma^+ \cap \Omega^+)} + \mu^{-n} \left\| \frac{f^-}{D^*} \right\|^n_{L^n(\Gamma^+ \cap \Omega^+)} \right] \right\} - 1 \right\}.$$

If $f \not\equiv 0$, we choose $\mu = \|\frac{f^-}{D^*}\|_{L^n(\Gamma^+ \cap \Omega^+)}$. If $f \not\equiv 0$, we may choose any $\mu > 0$ and then let $\mu \to 0$. □

In what follows we use Theorem 2.21 and Lemma 2.24 to derive some a priori estimates for solutions to quasi-linear equations and fully nonlinear equations. In the next result we do not assume uniform ellipticity.

PROPOSITION 2.27 *Suppose that $u \in C(\bar{\Omega}) \cap C^2(\Omega)$ satisfies*

$$Qu \equiv a_{ij}(x,u,Du)D_{ij}u + b(x,u,Du) = 0 \quad \text{in } \Omega$$

where $a_{ij} \in C(\Omega \times \mathbb{R} \times \mathbb{R}^n)$ satisfies

$$a_{ij}(x,z,p)\xi_i\xi_j > 0 \quad \text{for any } (x,z,p) \in \Omega \times \mathbb{R} \times \mathbb{R}^n \text{ and } \xi \in \mathbb{R}^n.$$

Suppose there exist nonnegative functions $g \in L^n_{\mathrm{loc}}(\mathbb{R}^n)$ and $h \in L^n(\Omega)$ such that

$$\frac{|b(x,z,p)|}{nD^*} \le \frac{h(x)}{g(p)} \quad \text{for any } (x,z,p) \in \Omega \times \mathbb{R} \times \mathbb{R}^n,$$

$$\int_\Omega h^n(x)dx < \int_{\mathbb{R}^n} g^n(p)dp \equiv g_\infty.$$

Then there holds $\sup_\Omega |u| \le \sup_{\partial\Omega} |u| + C \operatorname{diam}(\Omega)$ where C is a positive constant depending only on g and h.

EXAMPLE. The prescribed mean curvature equation is given by

$$(1+|Du|^2)\Delta u - D_iuD_juD_{ij}u = nH(x)(1+|Du|^2)^{3/2}$$

for some $H \in C(\Omega)$. We have

$$a_{ij}(x,z,p) = (1+|p|^2)\delta_{ij} - p_ip_j \Rightarrow D = (1+|p|^2)^{n-1}$$
$$b = -nH(x)(1+|p|^2)^{3/2}.$$

This implies

$$\frac{|b(x,z,p)|}{nD^*} \le \frac{|H(x)|(1+|p|^2)^{3/2}}{(1+|p|^2)^{\frac{n-1}{n}}} = |H(x)|(1+|p|^2)^{\frac{n+2}{2n}}$$

and in particular

$$g_\infty = \int_{\mathbb{R}^n} g^n(p)dp = \int_{\mathbb{R}^n} \frac{d}{p}(1+|p|^2)^{\frac{n+2}{2}} = \omega_n.$$

COROLLARY 2.28 *Suppose $u \in C(\bar{\Omega}) \cap C^2(\Omega)$ satisfies*

$$(1+|Du|^2)\Delta u - D_iuD_juD_{ij}u = nH(x)(1+|Du|^2)^{3/2} \quad \text{in } \Omega$$

for some $H \in C(\Omega)$. Then if

$$H_0 \equiv \int_\Omega |H(x)|^n \, dx < \omega_n,$$

we have

$$\sup_\Omega |u| \le \sup_{\partial\Omega} |u| + C \operatorname{diam}(\Omega)$$

where C is a positive constant depending only on n and H_0.

PROOF OF PROPOSITION 2.27: We prove for subsolutions. Assume $Qu \geq 0$ in Ω. Then we have

$$-a_{ij} D_{ij} u \leq b \quad \text{in } \Omega.$$

Note that $\{D_{ij}u\}$ is nonpositive in Γ^+. Hence $-a_{ij} D_{ij} u \geq 0$, which implies $b(x, u, Du) \geq 0$ in Γ^+. Then in $\Gamma^+ \cap \Omega^+$ there holds

$$\frac{b(x, z, Du)}{nD^*} \leq \frac{h(x)}{g(Du)}.$$

We may apply Lemma 2.24 to g^n and get

$$\begin{aligned} \int_{B_{\widetilde{M}}(0)} g^n &\leq \int_{\Gamma^+ \cap \Omega^+} g^n(Du) \left(\frac{-a_{ij} D_{ij} u}{nD^*} \right)^n \\ &\leq \int_{\Gamma^+ \cap \Omega^+} g^n(Du) \left(\frac{b}{nD^*} \right)^n \leq \int_{\Gamma^+ \cap \Omega^+} h^n \leq \int_{\Omega} h^n \left(< \int_{\mathbb{R}^n} g^n \right). \end{aligned}$$

Therefore there exists a positive constant C, depending only on g and h, such that $\widetilde{M} \leq C$. This implies

$$\sup_{\Omega} u \leq \sup_{\partial\Omega} u^+ + C \operatorname{diam}(\Omega).$$

□

Next we discuss Monge-Ampère equations.

COROLLARY 2.29 *Suppose $u \in C(\bar{\Omega}) \cap C^2(\Omega)$ satisfies*

$$\det(D^2 u) = f(x, u, Du) \quad \text{in } \Omega$$

for some $f \in C(\Omega \times \mathbb{R} \times \mathbb{R}^n)$. Suppose there exist nonnegative functions $g \in L^1_{\text{loc}}(\mathbb{R}^n)$ and $h \in L^1(\Omega)$ such that

$$|f(x, z, p)| \leq \frac{h(x)}{g(p)} \quad \text{for any } (x, z, p) \in \Omega \times \mathbb{R} \times \mathbb{R}^n,$$

$$\int_{\Omega} h(x)dx < \int_{\mathbb{R}^n} g(p)dp \equiv g_\infty.$$

Then there holds

$$\sup_{\Omega} |u| \leq \sup_{\partial\Omega} |u| + C \operatorname{diam}(\Omega)$$

where C is a positive constant depending only on g and h.

The proof is similar to that of Proposition 2.27. There are two special cases. The first case is given by $f = f(x)$. We may take $g \equiv 1$ and hence $g_\infty = \infty$. So we obtain the following:

COROLLARY 2.30 *Let $u \in C(\bar{\Omega}) \cap C^2(\Omega)$ satisfy*

$$\det(D^2 u) = f(x) \quad \text{in } \Omega$$

for some $f \in C(\bar{\Omega})$. Then there holds

$$\sup_{\Omega} |u| \le \sup_{\partial\Omega} |u| + \frac{\operatorname{diam}(\Omega)}{\omega_n^{1/n}} \left(\int_{\Omega} |f|^n \right)^{1/n}.$$

The second case is about the prescribed Gaussian curvature equations.

COROLLARY 2.31 *Let $u \in C(\bar{\Omega}) \cap C^2(\Omega)$ satisfy*

$$\det(D^2 u) = K(x)(1 + |Du|^2)^{\frac{n+2}{2}} \quad \text{in } \Omega$$

for some $K \in C(\bar{\Omega})$. Then if

$$K_0 \equiv \int_{\Omega} |K(x)| < \omega_n$$

we have

$$\sup_{\Omega} |u| \le \sup_{\partial\Omega} |u| + C \operatorname{diam}(\Omega)$$

where C is a positive constant depending only on n and K_0.

We finish this section by proving a maximum principle in a domain with small volume that is due to Varadhan.

Consider

$$Lu \equiv a_{ij} D_{ij} u + b_i D_i u + cu \quad \text{in } \Omega$$

where $\{a_{ij}\}$ is positive definite pointwise in Ω and

$$|b_i| + |c| \le \Lambda \quad \text{and} \quad \det(a_{ij}) \ge \lambda$$

for some positive constants λ and Λ.

THEOREM 2.32 *Suppose $u \in C(\bar{\Omega}) \cap C^2(\Omega)$ satisfies $Lu \ge 0$ in Ω with $u \le 0$ on $\partial\Omega$. Assume* $\operatorname{diam}(\Omega) \le d$. *Then there is a positive constant $\delta = \delta(n, \lambda, \Lambda, d) > 0$ such that if $|\Omega| \le \delta$ then $u \le 0$ in Ω.*

PROOF: If $c \le 0$, then $u \le 0$ by Theorem 2.21. In general, write $c = c^+ - c^-$. Then

$$a_{ij} D_{ij} u + b_i D_i u - c^- u \ge -c^+ u \ (\equiv f).$$

By Theorem 2.21 we have

$$\begin{aligned} \sup_{\Omega} u &\le c(n, \lambda, \Lambda, d) \|c^+ u^+\|_{L^n(\Omega)} \\ &\le c(n, \lambda, \Lambda, d) \|c^+\|_{L^\infty} |\Omega|^{1/n} \cdot \sup_{\Omega} u \le \tfrac{1}{2} \sup_{\Omega} u \end{aligned}$$

if $|\Omega|$ is small. Hence we get $u \le 0$ in Ω. □

REMARK 2.33. Compare this with Proposition 2.13, the maximum principle for a narrow domain.

2.6. Moving Plane Method

In this section we will use the moving plane method to discuss the symmetry of solutions. The following result was first proved by Gidas, Ni, and Nirenberg.

THEOREM 2.34 *Suppose $u \in C(\bar{B}_1) \cap C^2(B_1)$ is a positive solution of*

$$\begin{aligned} \Delta u + f(u) &= 0 \quad \text{in } B_1, \\ u &= 0 \quad \text{on } \partial B_1, \end{aligned}$$

where f is locally Lipschitz in $\mathbb{R}$. Then u is radially symmetric in B_1 and $\frac{\partial u}{\partial r}(x) < 0$ for $x \neq 0$.

The original proof requires that solutions be C^2 up to the boundary. Here we give a method that does not depend on the smoothness of domains nor the smoothness of solutions up to the boundary.

LEMMA 2.35 *Suppose that Ω is a bounded domain that is convex in the x_1-direction and symmetric with respect to the plane $\{x_1 = 0\}$. Suppose $u \in C(\bar{\Omega}) \cap C^2(\Omega)$ is a positive solution of*

$$\begin{aligned} \Delta u + f(u) &= 0 \quad \text{in } \Omega, \\ 8u &= 0 \quad \text{on } \partial\Omega, \end{aligned}$$

where f is locally Lipschitz in $\mathbb{R}$. Then u is symmetric with respect to x_1 and $D_{x_1}u(x) < 0$ for any $x \in \Omega$ with $x_1 > 0$.

PROOF: Write $x = (x_1, y) \in \Omega$ for $y \in \mathbb{R}^{n-1}$. We will prove

$$u(x_1, y) < u(x_1^*, y) \tag{2.4}$$

for any $x_1 > 0$ and $x_1^* < x_1$ with $x_1^* + x_1 > 0$. Then by letting $x_1^* \to -x_1$, we get $u(x_1, y) \leq u(-x_1, y)$ for any x_1. Then by changing the direction $x_1 \to -x_1$, we get the symmetry.

Let $a = \sup x_1$ for $(x_1, y) \in \Omega$. For $0 < \lambda < a$, define

$$\begin{aligned} \Sigma_\lambda &= \{x \in \Omega : x_1 > \lambda\}, \\ T_\lambda &= \{x_1 = \lambda\}, \\ \Sigma'_\lambda &= \text{reflection of } \Sigma_\lambda \text{ with respect to } T_\lambda, \\ x_\lambda &= (2\lambda - x_1, \dots, x_n) \quad \text{for } x = (x_1, \dots, x_n). \end{aligned}$$

In Σ_λ we define

$$w_\lambda(x) = u(x) - u(x_\lambda) \quad \text{for } x \in \Sigma_\lambda.$$

Then we have by the mean value theorem

$$\begin{aligned} \Delta w_\lambda + c(x, \lambda) w_\lambda &= 0 \quad \text{in } \Sigma_\lambda, \\ w_\lambda \leq 0 \text{ and } w_\lambda &\not\equiv 0 \quad \text{on } \partial\Sigma_\lambda, \end{aligned}$$

where $c(x, \lambda)$ is a bounded function in Σ_λ.

We need to show $w_\lambda < 0$ in Σ_λ for any $\lambda \in (0, a)$. This implies in particular that w_λ assumes along $\partial\Sigma_\lambda \cap \Omega$ its maximum in Σ_λ. By Theorem 2.5 (the Hopf lemma) we have for any such $\lambda \in (0, a)$

$$D_{x_1} w_\lambda \big|_{x_1=\lambda} = 2D_{x_1} u \big|_{x_1=\lambda} < 0.$$

For any λ close to a, we have $w_\lambda < 0$ by Proposition 2.13 (the maximum principle for a narrow domain) or Theorem 2.32. Let (λ_0, a) be the largest interval of values of λ such that $w_\lambda < 0$ in Σ_λ. We want to show $\lambda_0 = 0$. If $\lambda_0 > 0$, by continuity, $w_{\lambda_0} \le 0$ in Σ_{λ_0} and $w_{\lambda_0} \not\equiv 0$ on $\partial\Sigma_{\lambda_0}$. Then Theorem 2.7 (the strong maximum principle) implies $w_{\lambda_0} < 0$ in Σ_{λ_0}. We will show that for any small $\varepsilon > 0$

$$w_{\lambda_0-\varepsilon} < 0 \quad \text{in } \Sigma_{\lambda_0-\varepsilon}.$$

Fix $\delta > 0$ (to be determined). Let K be a closed subset in Σ_{λ_0} such that $|\Sigma_{\lambda_0} \setminus K| < \frac{\delta}{2}$. The fact that $w_{\lambda_0} < 0$ in Σ_{λ_0} implies

$$w_{\lambda_0}(x) \le -\eta < 0 \quad \text{for any } x \in K.$$

By continuity we have

$$w_{\lambda_0-\varepsilon} < 0 \quad \text{in } K.$$

For $\varepsilon > 0$ small, $|\Sigma_{\lambda_0-\varepsilon} \setminus K| < \delta$. We choose δ in such a way that we may apply Theorem 2.32 (the maximum principle for a domain with small volume) to $w_{\lambda_0-\varepsilon}$ in $\Sigma_{\lambda_0-\varepsilon} \setminus K$. Hence we get

$$w_{\lambda_0-\varepsilon}(x) \le 0 \quad \text{in } \Sigma_{\lambda_0-\varepsilon} \setminus K$$

and then by Theorem 2.10

$$w_{\lambda_0-\varepsilon}(x) < 0 \quad \text{in } \Sigma_{\lambda_0-\varepsilon} \setminus K.$$

Therefore we obtain for any small $\varepsilon > 0$

$$w_{\lambda_0-\varepsilon}(x) < 0 \quad \text{in } \Sigma_{\lambda_0-\varepsilon}.$$

This contradicts the choice of λ_0. □

CHAPTER 3

Weak Solutions: Part I

3.1. Guide

The goal of this chapter and the next is to discuss various regularity results for weak solutions to elliptic equations of divergence form. In order to explain ideas clearly we will discuss the equations with the following form only:

$$-D_j(a_{ij}(x)D_i u) + c(x)u = f(x).$$

We assume that Ω is a domain in $\mathbb{R}^n$. The function $u \in H^1(\Omega)$ is a *weak solution* if it satisfies

$$\int_\Omega (a_{ij} D_i u D_j \varphi + cu\varphi) = \int_\Omega f\varphi \quad \text{for any } \varphi \in H_0^1(\Omega),$$

where we assume

(i) the leading coefficients $a_{ij} \in L^\infty(\Omega)$ are *uniformly elliptic*, that is, for some positive constant λ there holds

$$a_{ij}(x)\xi_i\xi_j \geq \lambda|\xi|^2 \quad \text{for any } x \in \Omega \text{ and } \xi \in \mathbb{R}^n,$$

(ii) the coefficient $c \in L^{n/2}(\Omega)$ and nonhomogeneous term $f \in L^{\frac{2n}{n+2}}(\Omega)$.

Note by the Sobolev embedding theorem (ii) is the least assumption on c and f to have a meaningful equation.

We will prove various interior regularity results concerning the solution u if we have better assumptions on coefficients a_{ij} and c and the nonhomogeneous term f. Basically there are two class of regularity results, perturbation results and nonperturbation results. The first is based on the regularity assumption on the leading coefficients a_{ij}, which are assumed to be at least continuous. Under such an assumption we may compare solutions to the underlying equations with harmonic functions, or solutions to constant-coefficient equations. Then the regularity of solutions depends on how close they are to harmonic functions or how close the leading coefficients a_{ij} are to constant coefficients. In this direction we have Schauder estimates and $W^{2,p}$-estimates. In this chapter we only discuss the Schauder estimates.

For the second kind of regularity result, there is no continuity assumption on the leading coefficients a_{ij}. Hence the result is not based on the perturbation. The iteration methods introduced by De Giorgi and Moser are successful in dealing with nonperturbation situations. The results proved by them are fundamental to the discussion of quasi-linear equations, where the coefficients depend on the solutions. In fact, the linearity has no bearing in their arguments. This permits an extension of these results to quasi-linear equations with appropriate structure conditions. One may discuss boundary regularities in a similar way. We leave the details to the reader.

The outline of this chapter is as follows: The first section provides some general knowledge of Campanato and BMO spaces that are needed in both Chapters 3 and 4. Sections 3.3 and 3.4 as well as Sections 5.4 and 5.5 can be viewed as perturbation theory (from the constant-coefficient equations). The former deals with equations of divergence type, and the latter is for nondivergence type equations. The classical theory of Schauder estimates and L^p-estimates are also contained in the latter treatment. Note we did not use the classical potential estimates. Here two papers by Caffarelli [**2, 3**] and the book of Giaquinta [**5**] are sources for further reading.

3.2. Growth of Local Integrals

Let $B_R(x_0)$ be the ball in $\mathbb{R}^n$ of radius R centered at x_0. The well-known Sobolev theorem states that if $u \in W^{1,p}(B_R(x_0))$ with $p > n$ then u is Hölder-continuous with exponent $\alpha = 1 - n/p$.

In the first part of this section we prove a general result, due to S. Campanato, which characterizes Hölder-continuous functions by the growth of their local integrals. This result will be very useful for studying the regularity of solutions to elliptic differential equations. In the second part of this section we prove a result, due to John and Nirenberg, which gives an equivalent definition of functions of bounded mean oscillation.

Let Ω be a bounded connected open set in $\mathbb{R}^n$ and let $u \in L^1(\Omega)$. For any ball $B_r(x_0) \subset \Omega$, define

$$u_{x_0,r} = \frac{1}{|B_r(x_0)|} \int_{B_r(x_0)} u.$$

THEOREM 3.1 *Suppose $u \in L^2(\Omega)$ satisfies*

$$\int_{B_r(x)} |u - u_{x,r}|^2 \le M^2 r^{n+2\alpha} \quad \text{for any } B_r(x) \subset \Omega$$

for some $\alpha \in (0,1)$. Then $u \in C^\alpha(\Omega)$ and for any $\Omega' \Subset \Omega$ there holds

$$\sup_{\Omega'} |u| + \sup_{\substack{x,y \in \Omega' \\ x \neq y}} \frac{|u(x) - u(y)|}{|x - y|^\alpha} \le c\{M + \|u\|_{L^2(\Omega)}\}$$

where $c = c(n, \alpha, \Omega, \Omega') > 0$.

PROOF: Denote $R_0 = \operatorname{dist}(\Omega', \partial\Omega)$. For any $x_0 \in \Omega'$ and $0 < r_1 < r_2 \le R_0$, we have

$$|u_{x_0,r_1} - u_{x_0,r_2}|^2 \le 2(|u(x) - u_{x_0,r_1}|^2 + |u(x) - u_{x_0,r_2}|^2)$$

and integrating with respect to x in $B_{r_1}(x_0)$

$$|u_{x_0,r_1} - u_{x_0,r_2}|^2 \leq \frac{2}{\omega_n r_1^n}\Big\{ \int_{B_{r_1}(x_0)} |u - u_{x_0,r_1}|^2 + \int_{B_{r_2}(x_0)} |u - u_{x_0,r_2}|^2 \Big\}$$

from which the estimate

$$|u_{x_0,r_1} - u_{x_0,r_2}|^2 \leq c(n) M^2 r_1^{-n} \{ r_1^{n+2\alpha} + r_2^{n+2\alpha} \} \tag{3.1}$$

follows.

For any $R \leq R_0$, with $r_1 = R/2^{i+1}, r_2 = R/2^i$, we obtain

$$|u_{x_0,2^{-(i+1)}R} - u_{x_0,2^{-i}R}| \leq c(n) 2^{-(i+1)\alpha} M R^\alpha$$

and therefore for $h < k$

$$|u_{x_0,2^{-h}R} - u_{x_0,2^{-k}R}| \leq \frac{c(n)}{2^{(h+1)\alpha}} M R^\alpha \sum_{i=0}^{k-h-1} \frac{1}{2^{i\alpha}} \leq \frac{c(n,\alpha)}{2^{h\alpha}} M R^\alpha .$$

This shows that $\{u_{x_0,2^{-i}R}\} \subset \mathbb{R}$ is a Cauchy sequence, hence a convergent one. Its limit $\widehat{u}(x_0)$ is independent of the choice of R, since (3.1) can be applied with $r_1 = 2^{-i} R$ and $r_2 = 2^{-i} \overline{R}$ whenever $0 < R < \overline{R} \leq R_0$. Thus we get

$$\widehat{u}(x_0) = \lim_{r \downarrow 0} u_{x_0,r}$$

with

$$|u_{x_0,r} - \widehat{u}(x_0)| \leq c(n,\alpha) M r^\alpha \tag{3.2}$$

for any $0 < r \leq R_0$.

Recall that $\{u_{x,r}\}$ converges, as $r \to 0+$, in $L^1(\Omega)$ to the function u, by the Lebesgue theorem, so we have $u = \widehat{u}$ a.e. and (3.2) implies that $\{u_{x,r}\}$ converges uniformly to $u(x)$ in Ω'. Since $x \mapsto u_{x,r}$ is continuous for any $r > 0$, $u(x)$ is continuous. By (3.2) we get

$$|u(x)| \leq C M R^\alpha + |u_{x,R}|$$

for any $x \in \Omega'$ and $R \leq R_0$. Hence u is bounded in Ω' with the estimate

$$\sup_{\Omega'} |u| \leq c\{ M R_0^\alpha + \|u\|_{L^2(\Omega)} \}.$$

Finally, we prove that u is Hölder-continuous. Let $x, y \in \Omega'$ with $R = |x - y| < R_0/2$. Then we have

$$|u(x) - u(y)| \leq |u(x) - u_{x,2R}| + |u(y) - u_{y,2R}| + |u_{x,2R} - u_{y,2R}|.$$

The first two terms on the right sides are estimated in (3.2). For the last term we write

$$|u_{x,2R} - u_{y,2R}| \leq |u_{x,2R} - u(\zeta)| + |u_{y,2R} - u(\zeta)|$$

and integrating with respect to ζ over $B_{2R}(x) \cap B_{2R}(y)$, which contains $B_R(x)$, yields

$$|u_{x,2R} - u_{y,2R}|^2 \le \frac{2}{|B_R(x)|}\left\{ \int_{B_{2R}(x)} |u - u_{x,2R}|^2 + \int_{B_{2R}(y)} |u - u_{y,2R}|^2 \right\}$$
$$\le c(n,\alpha) M^2 R^{2\alpha}.$$

Therefore we have

$$|u(x) - u(y)| \le c(n,\alpha) M |x-y|^\alpha.$$

For $|x-y| > R_0/2$ we obtain

$$|u(x) - u(y)| \le 2 \sup_{\Omega'} |u| \le c\left\{M + \frac{1}{R_0^\alpha}\|u\|_{L^2}\right\}|x-y|^\alpha.$$

This finishes the proof. □

A special case of the Sobolev theorem is an easy consequence of Theorem 3.1. In fact, we have the following result due to Morrey.

COROLLARY 3.2 *Suppose $u \in H^1_{\text{loc}}(\Omega)$ satisfies*

$$\int_{B_r(x)} |Du|^2 \le M^2 r^{n-2+2\alpha} \quad \textit{for any } B_r(x) \subset \Omega$$

for some $\alpha \in (0,1)$. Then $u \in C^\alpha(\Omega)$ and for any $\Omega' \Subset \Omega$ there holds

$$\sup_{\Omega'} |u| + \sup_{\substack{x,y\in\Omega' \\ x \ne y}} \frac{|u(x) - u(y)|}{|x-y|^\alpha} \le c\{M + \|u\|_{L^2(\Omega)}\}$$

where $c = c(n, \alpha, \Omega, \Omega') > 0$.

PROOF: By the Poincaré inequality, we obtain

$$\int_{B_r(x)} |u - u_{x,r}|^2 \le c(n) r^2 \int_{B_r(x)} |Du|^2 \le c(n) M^2 r^{n+2\alpha}.$$

By applying Theorem 3.1, we have the result. □

The following result will be needed in Section 3.3.

LEMMA 3.3 *Suppose $u \in H^1(\Omega)$ satisfies*

$$\int_{B_r(x_0)} |Du|^2 \le M r^\mu \quad \textit{for any } B_r(x_0) \subset \Omega$$

for some $\mu \in [0,n)$. Then for any $\Omega' \Subset \Omega$ there holds for any $B_r(x_0) \subset \Omega$ with $x_0 \in \Omega'$

$$\int_{B_r(x_0)} |u|^2 \le c(n, \lambda, \mu, \Omega, \Omega')\left\{M + \int_\Omega u^2\right\} r^\lambda$$

where $\lambda = \mu + 2$ if $\mu < n-2$ and λ is any number in $[0,n)$ if $n - 2 \le \mu < n$.

PROOF: As before, denote $R_0 = \mathrm{dist}(\Omega', \partial\Omega)$. For any $x_0 \in \Omega'$ and $0 < r \leq R_0$, the Poincaré inequality yields

$$\int_{B_r(x_0)} |u - u_{x_0,r}|^2 \leq cr^2 \int_{B_r(x_0)} |Du|^2\, dx \leq c(n)Mr^{\mu+2}.$$

This implies that

$$\int_{B_r(x_0)} |u - u_{x_0,r}|^2 \leq c(n)Mr^{\lambda}$$

where λ is as in Theorem 3.3. For any $0 < \rho < r \leq R_0$ we have

$$\begin{aligned} \int_{B_\rho(x_0)} u^2 &\leq 2 \int_{B_\rho(x_0)} |u_{x_0,r}|^2 + 2 \int_{B_\rho(x_0)} |u - u_{x_0,r}|^2 \\ &\leq c(n)\rho^n |u_{x_0,r}|^2 + 2 \int_{B_r(x_0)} |u - u_{x_0,r}|^2 \qquad (3.3) \\ &\leq c(n)\left(\frac{\rho}{r}\right)^n \int_{B_r(x_0)} u^2 + Mr^{\lambda} \end{aligned}$$

where we used

$$|u_{x_0,r}|^2 \leq \frac{c(n)}{r^n} \int_{B_r(x_0)} u^2.$$

Hence the function $\phi(r) = \int_{B_r(x_0)} u^2$ satisfies the inequality

$$\phi(\rho) \leq c(n)\left\{\left(\frac{\rho}{r}\right)^n \phi(r) + Mr^{\lambda}\right\} \quad \text{for any } 0 < \rho < r \leq R_0 \tag{3.4}$$

for some $\lambda \in (0, n)$. If we may replace the term Mr^{λ} in the right by $M\rho^{\lambda}$, we are done. In fact, we would obtain that for any $0 < \rho < r \leq R_0$ there holds

$$\int_{B_\rho(x_0)} u^2 \leq c\left\{\left(\frac{\rho}{r}\right)^{\lambda} \int_{B_r(x_0)} u^2 + M\rho^{\lambda}\right\}. \tag{3.5}$$

Choose $r = R_0$. This implies

$$\int_{B_\rho(x_0)} u^2 \leq c\rho^{\lambda}\left\{\int_{\Omega} u^2 + M\right\} \quad \text{for any } \rho \leq R_0.$$

□

For this purpose, we need the following technical lemma.

LEMMA 3.4 *Let $\phi(t)$ be a nonnegative and nondecreasing function on* $[0, R]$. *Suppose that*

$$\phi(\rho) \leq A\left[\left(\frac{\rho}{r}\right)^{\alpha} + \varepsilon\right]\phi(r) + Br^{\beta}$$

for any $0 < \rho \leq r \leq R$, *with* A, B, α, β *nonnegative constants and* $\beta < \alpha$. *Then for any* $\gamma \in (\beta, \alpha)$, *there exists a constant* $\varepsilon_0 = \varepsilon_0(A, \alpha, \beta, \gamma)$ *such that if* $\varepsilon < \varepsilon_0$ *we have for all* $0 < \rho \leq r \leq R$

$$\phi(\rho) \leq c\left\{\left(\frac{\rho}{r}\right)^{\gamma}\phi(r) + B\rho^{\beta}\right\}$$

where c *is a positive constant depending on* A, α, β, γ. *In particular, we have for any* $0 < r \leq R$

$$\phi(r) \leq c\left\{\frac{\phi(R)}{R^{\gamma}}r^{\gamma} + Br^{\beta}\right\}.$$

PROOF: For $\tau \in (0, 1)$ and $r < R$, we have

$$\phi(\tau r) \leq A\tau^{\alpha}[1 + \varepsilon\tau^{-\alpha}]\phi(r) + Br^{\beta}.$$

Choose $\tau < 1$ in such a way that $2A\tau^{\alpha} = \tau^{\gamma}$ and assume $\varepsilon_0\tau^{-\alpha} < 1$. Then we get for every $r < R$

$$\phi(\tau r) \leq \tau^{\gamma}\phi(r) + Br^{\beta}$$

and therefore for all integers $k > 0$

$$\phi(\tau^{k+1}r) \leq \tau^{\gamma}\phi(\tau^{k}r) + B\tau^{k\beta}r^{\beta} \leq \tau^{(k+1)\gamma}\phi(r) + B\tau^{k\beta}r^{\beta}\sum_{j=0}^{k}\tau^{j(\gamma-\beta)}$$

$$\leq \tau^{(k+1)\gamma}\phi(r) + \frac{B\tau^{k\beta}r^{\beta}}{1 - \tau^{\gamma-\beta}}.$$

By choosing k such that $\tau^{k+2}r < \rho \leq \tau^{k+1}r$, the last inequality gives

$$\phi(\rho) \leq \frac{1}{\tau^{\gamma}}\left(\frac{\rho}{r}\right)^{\gamma}\phi(r) + \frac{B\rho^{\beta}}{\tau^{2\beta}(1 - \tau^{\gamma-\beta})}.$$

□

In the rest of this section we discuss functions of bounded mean oscillation (BMO). The following result was proved by John and Nirenberg.

THEOREM 3.5 (John-Nirenberg Lemma) *Suppose* $u \in L^1(\Omega)$ *satisfies*

$$\int_{B_r(x)} |u - u_{x,r}| \leq Mr^n \quad \textit{for any } B_r(x) \subset \Omega.$$

Then there holds for any $B_r(x) \subset \Omega$

$$\int_{B_r(x)} e^{\frac{p_0}{M}|u-u_{x,r}|} \leq Cr^n$$

for some positive p_0 *and* C *depending only on* n.

REMARK 3.6. Functions satisfying the condition of Theorem 3.5 are called functions of bounded mean oscillation (BMO). We have the following relation:

$$L^\infty \subset_+ \text{BMO}.$$

The counterexample is given by the following function in $(0, 1) \subset \mathbb{R}$:

$$u(x) = \log(x).$$

For convenience we use cubes instead of balls. We need the Calderon-Zygmund decomposition. First we introduce some terminology.

Take the unit cube Q_0. Cut it into 2^n equally sized cubes, which we take as the first generation. Do the same cutting for these small cubes to get the second generation. Continue this process. These cubes (from all generations) are called *dyadic cubes*. Any $(k+1)$-generation cube Q comes from some k-generation cube $\tilde{Q}$, which is called the *predecessor* of Q.

LEMMA 3.7 *Suppose $f \in L^1(Q_0)$ is nonnegative and $\alpha > |Q_0|^{-1} \int_{Q_0} f$ is a fixed constant. Then there exists a sequence of (nonoverlapping) dyadic cubes $\{Q_j\}$ in Q_0 such that*

$$f(x) \le \alpha \text{ a.e. in } Q_0 \setminus \bigcup_j Q_j, \quad \alpha \le \frac{1}{|Q_j|} \int_{Q_j} f\, dx < 2^n \alpha.$$

PROOF: Cut Q_0 into 2^n dyadic cubes and keep the cube Q if $|Q|^{-1} \int_Q f \ge \alpha$. For others keep cutting and always keep the cube Q if $|Q|^{-1} \int_Q f \ge \alpha$ and cut the rest. Let $\{Q_j\}$ be the cubes we have kept during this infinite process. We only need to verify that

$$f(x) \le \alpha \quad \text{a.e. in } Q_0 \setminus \bigcup_j Q_j.$$

Indeed, any predecessor Q of Q_j that we have kept has to satisfy $\frac{1}{|Q|} \int_Q f\, dx < \alpha$. Thus for Q_j, one has $\alpha \le \frac{1}{|Q_j|} \int_{Q_j} f\, dx < 2^n \alpha$. Let $F = Q_0 \setminus \bigcup_j Q_j$. For any $x \in F$, from the way we collect $\{Q_j\}$, there exists a sequence of cubes Q^i containing x such that

$$\frac{1}{|Q^i|} \int_{Q^i} f < \alpha \quad \text{and} \quad \operatorname{diam}(Q^i) \to 0 \text{ as } i \to \infty.$$

By the Lebesgue density theorem this implies that

$$f \le \alpha \quad \text{a.e. in } F.$$

$\square$

PROOF OF THEOREM 3.5: Assume $\Omega = Q_0$. We may rewrite the assumption in terms of cubes as follows:

$$\int_Q |u - u_Q| < M|Q| \quad \text{for any } Q \subset Q_0.$$

We will prove that there exist two positive constants $c_1(n)$ and $c_2(n)$ such that for any $Q \subset Q_0$ there holds

$$|\{x \in Q : |u - u_Q| > t\}| \leq c_1 |Q| \exp\left(-\frac{c_2}{M} t\right).$$

Then Theorem 3.5 follows easily.

Assume without loss of generality $M = 1$. Choose

$$\alpha > 1 \geq |Q_0|^{-1} \int_{Q_0} |u - u_{Q_0}| dx.$$

Apply the Calderon-Zygmund decomposition to $f = |u - u_{Q_0}|$. There exists a sequence of (nonoverlapping) cubes $\{Q_j^{(1)}\}_{j=1}^{\infty}$ such that

$$\alpha \leq \frac{1}{|Q_j^{(1)}|} \int_{Q_j^{(1)}} |u - u_{Q_0}| < 2^n \alpha,$$

$$|u(x) - u_{Q_0}| \leq \alpha \quad \text{a.e. } x \in Q_0 \setminus \bigcup_{j=1}^{\infty} Q_j^{(1)}.$$

This implies

$$\sum_j |Q_j^{(1)}| \leq \frac{1}{\alpha} \int_{Q_0} |u - u_{Q_0}| \leq \frac{1}{\alpha} |Q_0|,$$

$$|u_{Q_j^{(1)}} - u_{Q_0}| \leq \frac{1}{|Q_j^{(1)}|} \int_{Q_j^{(1)}} |u - u_{Q_0}| dx \leq 2^n \alpha.$$

The definition of the BMO norm implies that for each j

$$\frac{1}{|Q_j^{(1)}|} \int_{Q_j^{(1)}} |u - u_{Q_j^{(1)}}| dx \leq 1 < \alpha.$$

Apply the decomposition procedure above to $f = |u - u_{Q_j^{(1)}}|$ in $Q_j^{(1)}$. There exists a sequence of (nonoverlapping) cubes $\{Q_j^{(2)}\}$ in $\bigcup_j Q_j^{(1)}$ such that

$$\sum_{j=1}^{\infty} |Q_j^{(2)}| \leq \frac{1}{\alpha} \sum_j \int_{Q_j^{(1)}} |u - u_{Q_j^{(1)}}| \leq \frac{1}{\alpha} \sum_j |Q_j^{(1)}| \leq \frac{1}{\alpha^2} |Q_0|$$

and

$$|u(x) - u_{Q_j^{(1)}}| \leq \alpha \quad \text{a.e. } x \in Q_j^{(1)} \setminus \bigcup Q_j^{(2)},$$

which implies

$$|u(x) - u_{Q_0}| \leq 2 \cdot 2^n \alpha \quad \text{a.e. } x \in Q_0 \setminus \bigcup_j Q_j^{(2)}.$$

Continue this process. For any integer $k \geq 1$ there exists a sequence of disjoint cubes $\{Q_j^{(k)}\}$ such that

$$\sum_j |Q_j^{(k)}| \leq \frac{1}{\alpha^k}|Q_0|$$

and

$$|u(x) - u_{Q_0}| \leq k2^n\alpha \quad \text{a.e. } x \in Q_0 \setminus \bigcup_j Q_j^{(k)}.$$

Thus

$$|\{x \in Q_0 : |u - u_{Q_0}| > 2^n k\alpha\}| \leq \sum_{j=1}^{\infty} |Q_j^{(k)}| \leq \frac{1}{\alpha^k}|Q_0|.$$

For any t there exists an integer k such that $t \in [2^n k\alpha, 2^n(k+1)\alpha)$. This implies

$$\alpha^{-k} = \alpha\alpha^{-(k+1)} = \alpha e^{-(k+1)\log\alpha} \leq \alpha e^{-\frac{\log\alpha}{2^n\alpha}t}.$$

This finishes the proof. □

3.3. Hölder Continuity of Solutions

In this section we will prove Hölder regularity for solutions. The basic idea is to freeze the leading coefficients and then to compare solutions with harmonic functions. The regularity of solutions depends on how close solutions are to harmonic functions. Hence we need some regularity assumption on the leading coefficients.

Suppose $a_{ij} \in L^\infty(B_1)$ is uniformly elliptic in $B_1 = B_1(0)$, that is,

$$\lambda|\xi|^2 \leq a_{ij}(x)\xi_i\xi_j \leq \Lambda|\xi|^2 \quad \text{for any } x \in B_1,\ \xi \in \mathbb{R}^n.$$

In the following we assume that a_{ij} is at least continuous. We assume that $u \in H^1(B_1)$ satisfies

$$(*) \qquad \int_{B_1} a_{ij}D_iuD_j\varphi + cu\varphi = \int_{B_1} f\varphi \quad \text{for any } \varphi \in H_0^1(B_1).$$

The main theorem we will prove are the following Hölder estimates for solutions.

THEOREM 3.8 *Let $u \in H^1(B_1)$ solve $(*)$. Assume $a_{ij} \in C^0(\bar{B}_1)$, $c \in L^n(B_1)$, and $f \in L^q(B_1)$ for some $q \in (\frac{n}{2}, n)$. Then $u \in C^\alpha(B_1)$ with $\alpha = 2 - \frac{n}{q} \in (0,1)$. Moreover, there exists an $R_0 = R_0(\lambda, \Lambda, \tau, \|c\|_{L^n})$ such that for any $x \in B_{1/2}$ and $r \leq R_0$ there holds*

$$\int_{B_r(x)} |Du|^2 \leq Cr^{n-2+2\alpha}\{\|f\|^2_{L^q(B_1)} + \|u\|^2_{H^1(B_1)}\}$$

where $C = C(\lambda, \Lambda, \tau, \|c\|_{L^n})$ is a positive constant with

$$|a_{ij}(x) - a_{ij}(y)| \leq \tau(|x - y|) \quad \textit{for any } x, y \in B_1.$$

REMARK 3.9. In the case where $c \equiv 0$, we may replace $\|u\|_{H^1(B_1)}$ with $\|Du\|_{L^2(B_1)}$.

The idea of the proof is to compare the solution u with harmonic functions and use the perturbation argument.

LEMMA 3.10 (Basic Estimates for Harmonic Functions) *Suppose* $\{a_{ij}\}$ *is a constant positive definite matrix with*

$$\lambda|\xi|^2 \leq a_{ij}\xi_i\xi_j \leq \Lambda|\xi|^2 \quad \text{for any } \xi \in \mathbb{R}^n$$

for some $0 < \lambda \leq \Lambda$. *Suppose* $w \in H^1(B_r(x_0))$ *is a weak solution of*

$$a_{ij}D_{ij}w = 0 \quad \text{in } B_r(x_0). \tag{3.6}$$

Then for any $0 < \rho \leq r$, *there hold*

$$\int_{B_\rho(x_0)} |Dw|^2 \leq c\left(\frac{\rho}{r}\right)^n \int_{B_r(x_0)} |Dw|^2,$$

$$\int_{B_\rho(x_0)} |Dw - (Dw)_{x_0,\rho}|^2 \leq c\left(\frac{\rho}{r}\right)^{n+2} \int_{B_r(x_0)} |Dw - (Dw)_{x_0,r}|^2,$$

where $c = c(\lambda, \Lambda)$.

PROOF: Note that if w is a solution of (3.6), so is any one of its derivatives. We may apply Lemma 1.41 to Dw. □

COROLLARY 3.11 (Comparison with Harmonic Functions) *Suppose* w *is as in Lemma* 3.10. *Then for any* $u \in H^1(B_r(x_0))$ *there hold for any* $0 < \rho \leq r$

$$\int_{B_\rho(x_0)} |Du|^2 \leq c\left\{\left(\frac{\rho}{r}\right)^n \int_{B_r(x_0)} |Du|^2 + \int_{B_r(x_0)} |D(u-w)|^2\right\}$$

and

$$\int_{B_\rho(x_0)} |Du - (Du)_{x_0,\rho}|^2 \leq \left\{\left(\frac{\rho}{r}\right)^{n+2} \int_{B_r(x_0)} |Du - (Du)_{x_0,r}|^2 + \int_{B_r(x_0)} |D(u-w)|^2\right\}$$

where c *is a positive constant depending only on* λ *and* Λ.

PROOF: We prove this by direct computation. In fact, with $v = u - w$ we have for any $0 < \rho \le r$

$$\begin{aligned}\int_{B_\rho(x_0)} |Du|^2 &\le 2 \int_{B_\rho(x_0)} |Dw|^2 + 2 \int_{B_\rho(x_0)} |Dv|^2 \\ &\le c\left(\frac{\rho}{r}\right)^n \int_{B_r(x_0)} |Dw|^2 + 2 \int_{B_r(x_0)} |Dv|^2 \\ &\le c\left(\frac{\rho}{r}\right)^n \int_{B_r(x_0)} |Du|^2 + c\left[1 + \left(\frac{\rho}{r}\right)^n\right] \int_{B_r(x_0)} |Dv|^2\end{aligned}$$

and

$$\begin{aligned}&\int_{B_\rho(x_0)} |Du - (Du)_{x_0,\rho}|^2 \\ &\quad\le 2 \int_{B_\rho(x_0)} |Du - (Dw)_{x_0,\rho}|^2 + 2 \int_{B_\rho(x_0)} |Dv|^2 \\ &\quad\le 4 \int_{B_\rho(x_0)} |Dw - (Dw)_{x_0,\rho}|^2 + 6 \int_{B_\rho(x_0)} |Dv|^2 \\ &\quad\le c\left(\frac{\rho}{r}\right)^{n+2} \int_{B_r(x_0)} |Dw - (Dw)_{x_0,r}|^2 + 6 \int_{B_r(x_0)} |Dv|^2 \\ &\quad\le c\left(\frac{\rho}{r}\right)^{n+2} \int_{B_r(x_0)} |Du - (Du)_{x_0,r}|^2 + c\left[1 + \left(\frac{\rho}{r}\right)^{n+2}\right] \int_{B_r(x_0)} |Dv|^2.\end{aligned}$$

□

REMARK 3.12. The regularity of u depends on how close u is to w, the solution to the constant-coefficient equation.

PROOF OF THEOREM 3.8: We shall decompose u into a sum $v + w$ where w satisfies a homogeneous equation and v has estimates in terms of nonhomogeneous terms.

For any $B_r(x_0) \subset B_1$ write the equation in the following form:

$$\int_{B_1} a_{ij}(x_0) D_i u D_j \varphi = \int_{B_1} f\varphi - cu\varphi + (a_{ij}(x_0) - a_{ij}(x)) D_i u D_j \varphi.$$

In $B_r(x_0)$ the Dirichlet problem

$$\int_{B_r(x_0)} a_{ij}(x_0) D_i w D_j \varphi = 0 \quad \text{for any } \varphi \in H_0^1(B_r(x_0))$$

has a unique solution with w with $w - u \in H_0^1(B_r(x_0))$. Obviously the function $v = u - w \in H_0^1(B_r(x_0))$ satisfies the equation

$$\int_{B_r(x_0)} a_{ij}(x_0) D_i v D_j \varphi = \int_{B_r(x_0)} f\varphi - cu\varphi + (a_{ij}(x_0) - a_{ij}(x)) D_i u D_j \varphi$$

$$\text{for any } \varphi \in H_0^1(B_r(x_0)).$$

By taking the test function $\varphi = v$ we obtain

$$\int_{B_r(x_0)} |Dv|^2 \leq c\Big\{\tau^2(r) \int_{B_r(x_0)} |Du|^2 + \Big(\int_{B_r(x_0)} |c|^n\Big)^{2/n} \int_{B_r(x_0)} u^2 + \Big(\int_{B_r(x_0)} |f|^{\frac{2n}{n+2}}\Big)^{\frac{n+2}{n}}\Big\}$$

where we use the Sobolev inequality

$$\Big(\int_{B_r(x_0)} v^{\frac{2n}{n-2}}\Big)^{\frac{n-2}{2n}} \leq c(n)\Big(\int_{B_r(x_0)} |Dv|^2\Big)^{1/2}$$

for $v \in H_0^1(B_r(x_0))$. Therefore Corollary 3.11 implies for any $0 < \rho \leq r$

$$\begin{aligned} \int_{B_\rho(x_0)} |Du|^2 \leq c\Big\{\Big[\Big(\frac{\rho}{r}\Big)^n + \tau^2(r)\Big] \int_{B_r(x_0)} |Du|^2 \\ + \Big(\int_{B_r(x_0)} |c|^n\Big)^{2/n} \int_{B_r(x_0)} u^2 + \Big(\int_{B_r(x_0)} |f|^{\frac{2n}{n+2}}\Big)^{\frac{n+2}{n}}\Big\} \end{aligned} \tag{3.7}$$

where c is a positive constant depending only on λ and Λ. By Hölder inequality there holds

$$\Big(\int_{B_r(x_0)} |f|^{\frac{2n}{n+2}}\Big)^{\frac{n+2}{n}} \leq \Big(\int_{B_r(x_0)} |f|^q\Big)^{2/q} r^{n-2+2\alpha}$$

where $\alpha = 2 - \frac{n}{q} \in (0,1)$ if $\frac{n}{2} < q < n$. Hence (3.7) implies for any $B_r(x_0) \subset B_1$ and any $0 < \rho \leq r$

$$\begin{aligned} \int_{B_\rho(x_0)} |Du|^2 \leq C\Big\{\Big[\Big(\frac{\rho}{r}\Big)^n + \tau^2(r)\Big] \int_{B_r(x_o)} |Du|^2 + r^{n-2+2\alpha} \|f\|^2_{L^q(B_1)} \\ + \Big(\int_{B_r(x_0)} |c|^n\Big)^{2/n} \int_{B_r(x_0)} u^2\Big\}. \end{aligned}$$

CASE 1. $c \equiv 0$.

We have for any $B_r(x_0) \subset B_1$ and for any $0 < \rho \le r$

$$\int_{B_\rho(x_0)} |Du|^2 \le C\left\{\left[\left(\frac{\rho}{r}\right)^n + \tau^2(r)\right] \int_{B_r(x_0)} |Du|^2 + r^{n-2+2\alpha}\|f\|^2_{L^q(B_1)}\right\}.$$

Now the result would follow if in the above inequality we could write $\rho^{n-2+2\alpha}$ instead of $r^{n-2+2\alpha}$. This is in fact true and is stated in Lemma 3.4. By Lemma 3.4, there exists an $R_0 > 0$ such that for any $x_0 \in B_{1/2}$ and any $0 < \rho < r \le R_0$ we have

$$\int_{B_\rho(x_0)} |Du|^2 \le C\left\{\left(\frac{\rho}{r}\right)^{n-2+2\alpha} \int_{B_r(x_0)} |Du|^2 + \rho^{n-2+2\alpha}\|f\|^2_{L^q(B_1)}\right\}.$$

In particular, taking $r = R_0$ yields for any $\rho < R_0$

$$\int_{B_\rho(x_0)} |Du|^2 \le C\rho^{n-2+2\alpha}\left\{\int_{B_1} |Du|^2 + \|f\|^2_{L^q(B_1)}\right\}.$$

CASE 2. General case.

We have for any $B_r(x_0) \subset B_1$ and any $0 < \rho \le r$

$$\begin{aligned}\int_{B_\rho(x_0)} |Du|^2 \le C\left\{\left[\left(\frac{\rho}{r}\right)^n + \tau^2(r)\right] \int_{B_r(x_0)} |Du|^2 + r^{n-2+2\alpha}\chi(F)\right. \\ \left. + \int_{B_r(x_0)} u^2\right\}\end{aligned} \tag{3.8}$$

where $\chi(F) = \|f\|^2_{L^q(B_1)}$. We will prove for any $x_0 \in B_{1/2}$ and any $0 < \rho < r \le \frac{1}{2}$

$$\begin{aligned}\int_{B_\rho(x_0)} |Du|^2 \le C\left\{\left[\left(\frac{\rho}{r}\right)^n + \tau^2(r)\right] \int_{B_r(x_0)} |Du|^2\right. \\ \left. + r^{n-2+2\alpha}\left[\chi(F) + \int_{B_1} u^2 + \int_{B_1} |Du|^2\right]\right\}.\end{aligned} \tag{3.9}$$

We need a bootstrap argument. First by Lemma 3.3, there exists an $R_1 \in (\frac{1}{2}, 1)$ such that there holds for any $x_0 \in B_{R_1}$ and any $0 < r \le 1 - R_1$

$$\int_{B_r(x_0)} u^2 \le Cr^{\delta_1}\left\{\int_{B_1} |Du|^2 + \int_{B_1} u^2\right\} \tag{3.10}$$

where $\delta_1 = 2$ if $n > 2$ and δ_1 is arbitrary in (0,2) if $n = 2$. This, with (3.8), yields

$$\int_{B_\rho(x_0)} |Du|^2 \leq$$

$$c\left\{\left[\left(\frac{\rho}{r}\right)^n + \tau^2(r)\right] \int_{B_r(x_0)} |Du|^2 + r^{n-2+2\alpha}\chi(F) + r^{\delta_1}\|u\|^2_{H^1(B_1)}\right\}.$$

Then (3.9) holds in the following cases:

(i) $n = 2$, by choosing $\delta_1 = 2\alpha$;
(ii) $n > 2$ while $n - 2 + 2\alpha \leq 2$, by choosing $\delta_1 = 2$.

For $n > 2$ and $n - 2 + 2\alpha > 2$, we have

$$\int_{B_\rho(x_0)} |Du|^2 \leq c\left\{\left[\left(\frac{\rho}{r}\right)^n + \tau^2(r)\right] \int_{B_r(x_0)} |Du|^2 + r^2[\chi(F) + \|u\|^2_{H^1(B_1)}]\right\}.$$

Lemma 3.4 again yields for any $x_0 \in B_{R_1}$ and any $0 < r \leq 1 - R_1$

$$\int_{B_r(x_0)} |Du|^2 \leq Cr^2\{\chi(F) + \|u\|^2_{H^1(B_1)}\}.$$

Hence by Lemma 3.3, there exists an $R_2 \in (\frac{1}{2}, R_1)$ such that there holds for any $x_0 \in B_{R_2}$ and any $0 < r \leq R_1 - R_2$

$$\int_{B_r(x_0)} u^2 \leq Cr^{\delta_2}\{\chi(F) + \|u\|^2_{H^1(B_1)}\} \tag{3.11}$$

where $\delta_2 = 4$ if $n > 4$ and δ_2 is arbitrary in $(2, n)$ if $n = 3$ or 4. Notice (3.11) is an improvement compared with (3.10). Substitute (3.11) in (3.8) and continue the process. After finite steps, we get (3.9). This finishes the proof. □

3.4. Hölder Continuity of Gradients

In this section we will prove Hölder regularity for gradients of solutions. We follow the same idea used to prove Theorem 3.8.

Suppose $a_{ij} \in L^\infty(B_1)$ is uniformly elliptic in $B_1 = B_1(0)$, that is,

$$\lambda|\xi|^2 \leq a_{ij}(x)\xi_i\xi_j \leq \Lambda|\xi|^2 \quad \text{for any } x \in B_1,\ \xi \in \mathbb{R}^n.$$

We assume that $u \in H^1(B_1)$ satisfies

$$\int_{B_1} a_{ij}D_iuD_j\varphi + cu\varphi = \int_{B_1} f\varphi \quad \text{for any } \varphi \in H_0^1(B_1). \tag{$*$}$$

The main theorems we will prove are the following Hölder estimates for gradients.

THEOREM 3.13 *Let $u \in H^1(B_1)$ solve $(*)$. Assume $a_{ij} \in C^\alpha(\bar{B}_1)$, $c \in L^q(B_1)$ and $f \in L^q(B_1)$ for some $q > n$ and $\alpha = 1 - \frac{n}{q} \in (0, 1)$. Then $Du \in C^\alpha(B_1)$. Moreover, there exists an $R_0 = R_0(\lambda, |a_{ij}|_{C^\alpha}, |c|_{L^q})$ such that for any $x \in B_{1/2}$ and $r \leq R_0$ there holds*

$$\int_{B_r(x)} |Du - (Du)_{x,r}|^2 \leq C r^{n+2\alpha} \{\|f\|^2_{L^q(B_1)} + \|u\|^2_{H^1(B_1)}\}$$

where $C = C(\lambda, |a_{ij}|_{C^\alpha}, |c|_{L^q})$ is a positive constant.

PROOF: We shall decompose u into a sum $v + w$ where w satisfies a homogeneous equation and v has estimates in terms of nonhomogeneous terms.

For any $B_r(x_0) \subset B_1$ write the equation in the following form:

$$\int_{B_1} a_{ij}(x_0) D_i u D_j \varphi = \int_{B_1} f\varphi - cu\varphi + (a_{ij}(x_0) - a_{ij}(x)) D_i u D_j \varphi.$$

In $B_r(x_0)$ the Dirichlet problem

$$\int_{B_r(x_0)} a_{ij}(x_0) D_i w D_j \varphi = 0 \quad \text{for any } \varphi \in H^1_0(B_r(x_0))$$

has a unique solution w with $w - u \in H^1_0(B_r(x_0))$. Obviously the function $v = u - w \in H^1_0(B_r(x_0))$ satisfies the equation

$$\int_{B_r(x_0)} a_{ij}(x_0) D_i v D_j \varphi = \int_{B_r(x_0)} f\varphi - cu\varphi + (a_{ij}(x_0) - a_{ij}(x)) D_i u D_j \varphi$$

$$\text{for any } \varphi \in H^1_0(B_r(x_0)).$$

By taking the test function $\varphi = v$ we obtain

$$\int_{B_r(x_0)} |Dv|^2 \leq c\Bigg\{\tau^2(r) \int_{B_r(x_0)} |Du|^2 + \Bigg(\int_{B_r(x_0)} |c|^n\Bigg)^{\frac{2}{n}} \int_{B_r(x_0)} u^2 + \Bigg(\int_{B_r(x_0)} |f|^{\frac{2n}{n+2}}\Bigg)^{\frac{n+2}{n}}\Bigg\}.$$

Therefore Corollary 3.11 implies for any $0 < \rho \leq r$

$$\int_{B_\rho(x_0)} |Du|^2 \leq c\Bigg\{\Bigg[\Big(\frac{\rho}{r}\Big)^n + \tau^2(r)\Bigg] \int_{B_r(x_0)} |Du|^2 + \Bigg(\int_{B_r(x_0)} |c|^n\Bigg)^{\frac{2}{n}} \int_{B_r(x_0)} u^2 + \Bigg(\int_{B_r(x_0)} |f|^{\frac{2n}{n+2}}\Bigg)^{\frac{n+2}{n}}\Bigg\} \tag{3.12}$$

and

$$\int_{B_\rho(x_0)} |Du - (Du)_{x_0,\rho}|^2$$

$$\leq c\left\{\left(\frac{\rho}{r}\right)^{n+2} \int_{B_r(x_0)} |Du - (Du)_{x_0,r}|^2 + \tau^2(r) \int_{B_r(x_0)} |Du|^2 \right. \tag{3.13}$$

$$\left. + \left(\int_{B_r(x_0)} |c|^n\right)^{\frac{2}{n}} \int_{B_r(x_0)} u^2 + \left(\int_{B_r(x_0)} |f|^{\frac{2n}{n+2}}\right)^{\frac{n+2}{n}}\right\},$$

where c is a positive constant depending only on λ and Λ.

By the Hölder inequality we have for any $B_r(x_0) \subset B_1$

$$\left(\int_{B_r(x_0)} |f|^{\frac{2n}{n+2}}\right)^{\frac{n+2}{n}} \leq \left(\int_{B_r(x_0)} |f|^q\right)^{\frac{2}{q}} r^{n+2\alpha},$$

$$\left(\int_{B_r(x_0)} |c|^n\right)^{\frac{2}{n}} \leq r^{2\alpha}\left(\int_{B_r(x_0)} |c|^q\right)^{\frac{2}{q}}$$

with $\alpha = 1 - \frac{n}{q}$.

CASE 1. $a_{ij} \equiv \text{const}$, $c \equiv 0$.

In this case $\tau(r) \equiv 0$. Hence by estimate (3.13) there holds for any $B_r(x_0) \subset B_1$ and $0 < \rho \leq r$,

$$\int_{B_\rho(x_0)} |Du - (Du)_{x_0,\rho}|^2 \leq$$

$$c\left\{\left(\frac{\rho}{r}\right)^{n+2} \int_{B_r(x_0)} |Du - (Du)_{x_0,r}|^2 + r^{n+2\alpha}\|f\|^2_{L^q(B_1)}\right\}.$$

By Lemma 1.4, we may replace $r^{n+2\alpha}$ by $\rho^{n+2\alpha}$ to get the result.

CASE 2. $c \equiv 0$.

By (3.12) and (3.13), we have for any $B_r(x_0) \subset B_1$ and any $\rho < r$

$$\int_{B_\rho(x_0)} |Du|^2 \leq c\left\{\left[\left(\frac{\rho}{r}\right)^n + r^{2\alpha}\right] \int_{B_r(x_0)} |Du|^2 + r^{n+2\alpha}\|f\|^2_{L^q(B_1)}\right\} \tag{3.14}$$

and

$$\int_{B_\rho(x_0)} |Du - (Du)_{x_0,\rho}|^2$$

$$\leq C\left\{\left(\frac{\rho}{r}\right)^{n+2} \int_{B_r(x_0)} |Du - (Du)_{x_0,r}|^2 + r^{2\alpha} \int_{B_r(x_0)} |Du|^2 + r^{n+2\alpha}\|f\|^2_{L^q(B_1)}\right\}. \tag{3.15}$$

We need to estimate the integral $\int_{B_r(x_0)} |Du|^2$. Write $\chi(F) = \|f\|^2_{L^q(B_1)}$. Take small $\delta > 0$. Then (3.14) implies

$$\int_{B_\rho(x_0)} |Du|^2 \leq C\left\{\left[\left(\frac{\rho}{r}\right)^n + r^{2\alpha}\right] \int_{B_r(x_0)} |Du|^2 + r^{n-2\delta}\chi(F)\right\}.$$

Hence Lemma 3.4 implies the existence of an $R_1 \in (\frac{3}{4}, 1)$ with $r_1 = 1 - R_1$ such that for any $x_0 \in B_{R_1}$ and any $0 < r \leq r_1$ there holds

$$\int_{B_r(x_0)} |Du|^2 \leq Cr^{n-2\delta}\{\chi(F) + \|Du\|^2_{L^2(B_1)}\}. \tag{3.16}$$

Therefore by substituting (3.16) in (3.15) we obtain for any $0 < \rho < r \leq r_1$

$$\int_{B_\rho(x_0)} |Du - (Du)_{x_0,\rho}|^2 \leq$$

$$C\left\{\left(\frac{\rho}{r}\right)^{n+2} \int_{B_r(x_0)} |Du - (Du)_{x_0,r}|^2 + r^{n+2\alpha-2\delta}[\chi(F) + \|Du\|^2_{L^2(B_1)}]\right\}.$$

By Lemma 3.4 again, there holds for any $x_0 \in B_{R_1}$ and any $0 < \rho < r \leq r_1$

$$\int_{B_\rho(x_0)} |Du - (Du)_{x_0,\rho}|^2$$

$$\leq C\left\{\left(\frac{\rho}{r}\right)^{n+2\alpha-2\delta} \int_{B_r} |Du - (Du)_{x_0,r}|^2 + \rho^{n+2\alpha-2\delta}[\chi(F) + \|Du\|^2_{L^2(B_1)}]\right\}.$$

With $r = r_1$ this implies that for any $x_0 \in B_{R_1}$ and any $0 < r \leq r_1$

$$\int_{B_r(x_0)} |Du - (Du)_{x_0,r}|^2 \leq Cr^{n+2\alpha-2\delta}\{\chi(F) + \|Du\|^2_{L^2(B_1)}\}.$$

Hence $Du \in C^{\alpha-\delta}_{\rm loc}$ for any $\delta > 0$ small. In particular, $Du \in L^\infty_{\rm loc}$ and there holds

$$\sup_{B_{3/4}} |Du|^2 \le C\{\chi(F) + \|Du\|^2_{L^2(B_1)}\}. \tag{3.17}$$

By combining (3.15) and (3.17), there holds for $0 < \rho < r \le r_1$ and any $x_0 \in B_{1/2}$

$$\int_{B_\rho(x_0)} |Du - (Du)_{x_0,\rho}|^2 \le \\ C\left\{\left(\frac{\rho}{r}\right)^{n+2} \int_{B_r(x_0)} |Du - (Du)_{x_0,r}|^2 + r^{n+2\alpha}[\chi(F) + \|Du\|^2_{L^2(B_1)}]\right\}.$$

By Lemma 3.4 again, this implies

$$\int_{B_\rho(x_0)} |Du - (Du)_{x_0,\rho}|^2 \le \\ C\left\{\left(\frac{\rho}{r}\right)^{n+2\alpha} \int_{B_r(x_0)} |Du - (Du)_{x_0,r}|^2 + \rho^{n+2\alpha}[\chi(F) + \|Du\|^2_{L^2(B_1)}]\right\}.$$

Choose $r = r_1$. We have for any $x_0 \in B_{1/2}$ and $r \le r_1$

$$\int_{B_r(x_0)} |Du - (Du)_{x_0,r}|^2 \le c r^{n+2\alpha}\{\chi(F) + \|Du\|^2_{L^2(B_1)}\}.$$

CASE 3. General case.

By (3.12) and (3.13) we have for any $B_r(x_0) \subset B_1$ and $\rho < r$

$$\int_{B_\rho(x_0)} |Du|^2 \le C\left\{\left[\left(\frac{\rho}{r}\right)^n + r^{2\alpha}\right] \int_{B_r(x_0)} |Du|^2 \\ + \int_{B_r(x_0)} u^2 + r^{n+2\alpha}\chi(F)\right\} \tag{3.18}$$

and

$$\int_{B_\rho(x_0)} |Du - (Du)_{x_0,\rho}|^2 \\ \le C\left\{\left(\frac{\rho}{r}\right)^{n+2} \int_{B_r(x_0)} |Du - (Du)_{x_0,r}|^2 \\ + r^{2\alpha}\left[\int_{B_r(x_0)} u^2 + \int_{B_r(x_0)} |Du|^2\right] + r^{n+2\alpha}\chi(F)\right\} \tag{3.19}$$

where $\chi(F) = \|f\|^2_{L^q(B_1)}$.

In (3.18), we may replace $r^{n+2\alpha}$ by r^n. As in the proof of Theorem 3.8, we can show that for any small $\delta > 0$ there exists an $R_1 \in (\frac{3}{4}, 1)$ such that for any $x \in B_{R_1}$ and $r < 1 - R_1$

$$\int_{B_r(x_0)} |Du|^2 \le C r^{n-2\delta}\{\chi(F) + \|u\|^2_{H^1(B)}\}. \tag{3.20}$$

By Lemma 3.3, we also get

$$\int_{B_r(x_0)} u^2 \le C r^{n-2\delta}\{\chi(F) + \|u\|^2_{H^1(B)}\}. \tag{3.21}$$

Write $\chi(F, u) = \|f\|^2_{L^q} + \|u\|^2_{H^1}$. Then (3.19), (3.20), and (3.21) imply that

$$\int_{B_\rho(x_0)} |Du - (Du)_{x_0,\rho}|^2 \le c\left\{\left(\frac{\rho}{r}\right)^{n+2} \int_{B_r(x_0)} |Du - (Du)_{x_0,r}|^2 + r^{n+2\alpha-2\delta}\chi(F, u)\right\}.$$

Hence Lemma 3.4 and Theorem 3.1 imply that $Du \in C^{\alpha-\delta}_{\mathrm{loc}}$ for small $\delta < \alpha$. In particular, $u \in C^1_{\mathrm{loc}}$ with the estimate

$$\sup_{B_{3/4}} |u|^2 + \sup_{B_{3/4}} |Du|^2 \le C\chi(F, u). \tag{3.22}$$

Now (3.19) and (3.22) imply that

$$\int_{B_\rho(x_0)} |Du - (Du)_{x_0,\rho}|^2 \le c\left\{\left(\frac{\rho}{r}\right)^{n+2} \int_{B_r(x_0)} |Du - (Du)_{x_0,r}|^2 + r^{n+2\alpha}\chi(F, u)\right\}.$$

This finishes the proof of Theorem 3.8. □

REMARK 3.14. It is natural to ask whether $f \in L^\infty(B_1)$, with appropriate assumptions on a_{ij} and c, implies $Du \in C^1_{\mathrm{loc}}$. Consider a special case

$$\int_{B_1} D_i u D_i \varphi = \int_{B_1} f\varphi \quad \text{for any } \varphi \in H^1_0(B_1).$$

There exists an example showing that $f \in C$ and $u \in C^{1,\alpha}_{\mathrm{loc}}$ for any $\alpha \in (0, 1)$ while $D^2 u \notin C$.

EXAMPLE. In the n-dimensional ball $B_R = B_R(0)$ of radius $R < 1$ consider

$$\Delta u = \frac{x_2^2 - x_1^2}{2|x|^2}\left\{\frac{n+2}{(-\log|x|)^{1/2}} + \frac{1}{2(-\log|x|)^{3/2}}\right\}$$

where the right side is continuous in $\overline{B}_R$ if we set it equal to 0 at the origin. The function $u(x) = (x_1^2 - x_2^2)(-\log|x|)^{1/2} \in C(\overline{B}_R) \cap C^\infty(\overline{B}_R \setminus \{0\})$ satisfies the above equation in $B_R \setminus \{0\}$ and the boundary condition $u = \sqrt{-\log R}(x_1^2 - x_2^2)$ on ∂B_R. But u cannot be a classical solution of the problem since $\lim_{|x|\to 0} D_{11}u = \infty$ and therefore u is not in $C^2(B_R)$. In fact, the problem has no classical solution (although it has a weak solution).

Assume on the contrary that a classical solution v exists. Then the function $w = u - v$ is harmonic and bounded in $B_R \setminus \{0\}$. By a theorem from harmonic function theory on removable singularities, w may be redefined at the origin so that $\Delta w = 0$ in B_R and therefore belongs to $C^2(B_R)$. In particular, the (finite) limit $\lim_{|x|\to 0} D_{11}u$ exists, which is a contradiction.

CHAPTER 4

Weak Solutions, Part II

4.1. Guide

This chapter covers the well-known theory of De Giorgi–Nash–Moser. We present the approaches of both De Giorgi and Moser so students can make comparisons and can see that the ideas involved are essentially the same. The classical paper [**12**] is certainly very nice material for further reading; one may also wish to compare the results in [**7, 12**].

4.2. Local Boundedness

In the following three sections we will discuss the De Giorgi–Nash–Moser theory for linear elliptic equations. In this section we will prove the local boundedness of solutions. In the next section we will prove Hölder continuity. Then in Section 4.4 we will discuss the Harnack inequality. For all results in these three sections there is no regularity assumption of coefficients.

The main theorem of this section is the following boundedness result.

THEOREM 4.1 *Suppose* $a_{ij} \in L^\infty(B_1)$ *and* $c \in L^q(B_1)$ *for some* $q > \frac{n}{2}$ *satisfy the following assumptions*:

$$a_{ij}(x)\xi_i\xi_j \geq \lambda|\xi|^2 \quad \text{for any } x \in B_1,\ \xi \in \mathbb{R}^n,$$

and

$$|a_{ij}|_{L^\infty} + \|c\|_{L^q} \leq \Lambda$$

for some positive constants λ *and* Λ. *Suppose that* $u \in H^1(B_1)$ *is a subsolution in the following sense*:

$$(*) \qquad \int_{B_1} a_{ij} D_i u D_j \varphi + cu\varphi \leq \int_{B_1} f\varphi \quad \text{for any } \varphi \in H_0^1(B_1) \text{ and } \varphi \geq 0 \text{ in } B_1.$$

If $f \in L^q(B_1)$, *then* $u^+ \in L^\infty_{\mathrm{loc}}(B_1)$. *Moreover, there holds for any* $\theta \in (0,1)$ *and any* $p > 0$

$$\sup_{B_\theta} u^+ \leq C\left\{\frac{1}{(1-\theta)^{n/p}}\|u^+\|_{L^p(B_1)} + \|f\|_{L^q(B_1)}\right\}$$

where $C = C(n, \lambda, \Lambda, p, q)$ *is a positive constant.*

In the following we use two approaches to prove this theorem, one by De Giorgi and the other by Moser.

PROOF: We first prove the theorem for $\theta = \frac{1}{2}$ and $p = 2$.

METHOD 1: DE GIORGI'S APPROACH: Consider $v = (u - k)^+$ for $k \geq 0$ and $\zeta \in C_0^1(B_1)$. Set $\varphi = v\zeta^2$ as the test function. Note $v = u - k$, $Dv = Du$ a.e. in $\{u > k\}$ and $v = 0$, $Dv = 0$ a.e. in $\{u \leq k\}$. Hence if we substitute such defined φ in $(*)$, we integrate in the set $\{u > k\}$.

By the Hölder inequality we have

$$\begin{aligned}\int a_{ij} D_i u D_j \varphi &= \int a_{ij} D_i u D_j v \zeta^2 + 2 a_{ij} D_i u D_j \zeta v \zeta \\ &\geq \lambda \int |Dv|^2 \zeta^2 - 2\Lambda \int |Dv||D\zeta| v\zeta \\ &\geq \frac{\lambda}{2} \int |Dv|^2 \zeta^2 - \frac{2\Lambda^2}{\lambda} \int |D\zeta|^2 v^2.\end{aligned}$$

Hence we obtain

$$\int |Dv|^2 \zeta^2 \leq C\left\{\int v^2 |D\zeta|^2 + \int |c| v^2 \zeta^2 + k^2 \int |c| \zeta^2 + \int |f| v \zeta^2\right\}$$

from which the estimate

$$\int |D(v\zeta)|^2 \leq C\left\{\int v^2 |D\zeta|^2 + \int |c| v^2 \zeta^2 + k^2 \int |c| \zeta^2 + \int |f| v \zeta^2\right\}$$

follows.

Recall the Sobolev inequality for $v\zeta \in H_0^1(B_1)$,

$$\left(\int_{B_1} (v\zeta)^{2^*}\right)^{\frac{2}{2^*}} \leq c(n) \int_{B_1} |D(v\zeta)|^2$$

where $2^* = 2n/(n-2)$ for $n > 2$ and $2^* > 2$ is arbitrary if $n = 2$. The Hölder inequality implies that with $\delta > 0$ small and $\zeta \leq 1$

$$\begin{aligned}\int |f| v \zeta^2 &\leq \left(\int |f|^q\right)^{\frac{1}{q}} \left(\int |v\zeta|^{2^*}\right)^{\frac{1}{2^*}} |\{v\zeta \neq 0\}|^{1 - \frac{1}{2^*} - \frac{1}{q}} \\ &\leq c(n) \|f\|_{L^q} \left(\int |D(v\zeta)|^2\right)^{\frac{1}{2}} |\{v\zeta \neq 0\}|^{\frac{1}{2} + \frac{1}{n} - \frac{1}{q}} \\ &\leq \delta \int |D(v\zeta)|^2 + c(n, \delta) \|f\|_{L^q}^2 |\{v\zeta \neq 0\}|^{1 + \frac{2}{n} - \frac{2}{q}}.\end{aligned}$$

Note $1 + \frac{2}{n} - \frac{2}{q} > 1 - \frac{1}{q}$ if $q > \frac{n}{2}$. Therefore we have the following estimate:

$$\int |D(v\zeta)|^2 \leq C\left\{\int v^2 |D\zeta|^2 + \int |c| v^2 \zeta^2 + k^2 \int |c| \zeta^2 + F^2 |\{v\zeta \neq 0\}|^{1 - \frac{1}{q}}\right\}$$

where $F = \|f\|_{L^q(B_1)}$.

We claim that there holds

$$\int |D(v\zeta)|^2 \leq C\left\{\int v^2 |D\zeta|^2 + (k^2 + F^2) |\{v\zeta \neq 0\}|^{1 - \frac{1}{q}}\right\} \tag{4.1}$$

if $|\{v\zeta \neq 0\}|$ is small.

It is obvious if $c \equiv 0$. In fact, in this special case there is no restriction on the set $\{v\zeta \neq 0\}$. In general, the Hölder inequality implies that

$$\int |c|v^2\zeta^2 \leq \Big(\int |c|^q\Big)^{\frac{1}{q}}\Big(\int (v\zeta)^{2^*}\Big)^{\frac{2}{2^*}} |\{v\zeta \neq 0\}|^{1-\frac{2}{2^*}-\frac{1}{q}}$$
$$\leq c(n)\int |D(v\zeta)|^2\Big(\int |c|^q\Big)^{\frac{1}{q}} |\{v\zeta \neq 0\}|^{\frac{2}{n}-\frac{1}{q}}$$

and

$$\int |c|\zeta^2 \leq \Big(\int |c|^q\Big)^{\frac{1}{q}} |\{v\zeta \neq 0\}|^{1-\frac{1}{q}}.$$

Therefore we have

$$\int |D(v\zeta)|^2 \leq C\Big\{\int v^2|D\zeta|^2 + \int |D(v\zeta)|^2|\{v\zeta \neq 0\}|^{\frac{2}{n}-\frac{1}{q}}$$
$$+ (k^2 + F^2)|\{v\zeta \neq 0\}|^{1-\frac{1}{q}}\Big\}.$$

This implies (4.1) if $|\{v\zeta \neq 0\}|$ is small. To continue, we obtain by the Sobolev inequality

$$\int (v\zeta)^2 \leq \Big(\int (v\zeta)^{2^*}\Big)^{\frac{2}{2^*}} |\{v\zeta \neq 0\}|^{1-\frac{2}{2^*}}$$
$$\leq c(n)\int |D(v\zeta)|^2|\{v\zeta \neq 0\}|^{\frac{2}{n}}.$$

Therefore we have

$$\int (v\zeta)^2 \leq C\Big\{\int v^2|D\zeta|^2|\{v\zeta \neq 0\}|^{\frac{2}{n}} + (k+F)^2|\{v\zeta \neq 0\}|^{1+\frac{2}{n}-\frac{1}{q}}\Big\}$$

if $|\{v\zeta \neq 0\}|$ is small. Hence there exists an $\varepsilon > 0$ such that

$$\int (v\zeta)^2 \leq C\Big\{\int v^2|D\zeta|^2|\{v\zeta \neq 0\}|^{\varepsilon} + (k+F)^2|\{v\zeta \neq 0\}|^{1+\varepsilon}\Big\}$$

if $|\{v\zeta \neq 0\}|$ is small. Choose the cutoff function in the following way. For any fixed $0 < r < R \leq 1$ choose $\zeta \in C_0^\infty(B_R)$ such that $\zeta \equiv 1$ in B_r and $0 \leq \zeta \leq 1$ and $|D\zeta| \leq 2(R-r)^{-1}$ in B_1. Set

$$A(k,r) = \{x \in B_r : u \geq k\}.$$

We conclude that for any $0 < r < R \leq 1$ and $k > 0$

$$\int_{A(k,r)} (u-k)^2 \leq \tag{4.2}$$
$$C\Big\{\frac{1}{(R-r)^2}|A(k,R)|^{\varepsilon}\int_{A(k,R)} (u-k)^2 + (k+F)^2|A(k,R)|^{1+\varepsilon}\Big\}$$

if $|A(k,R)|$ is small. Note

$$|A(k,R)| \le \frac{1}{k} \int_{A(k,R)} u^+ \le \frac{1}{k} \|u^+\|_{L^2}.$$

Hence (4.2) holds if $k \ge k_0 = C\|u^+\|_{L^2}$ for some large C depending only on λ and Λ.

Next we would show that there exists some $k = C(k_0 + F)$ such that

$$\int_{A(k,1/2)} (u-k)^2 = 0.$$

To continue we take any $h > k \ge k_0$ and any $0 < r < 1$. It is obvious that $A(k,r) \supset A(h,r)$. Hence we have

$$\int_{A(h,r)} (u-h)^2 \le \int_{A(k,r)} (u-k)^2$$

and

$$|A(h,r)| = |B_r \cap \{u - k > h - k\}| \le \frac{1}{(h-k)^2} \int_{A(k,r)} (u-k)^2.$$

Therefore by (4.2) we have for any $h > k \ge k_0$ and $\frac{1}{2} \le r < R \le 1$

$$\begin{aligned}\int_{A(h,r)} (u-h)^2 &\le C\left\{\frac{1}{(R-r)^2} \int_{A(h,R)} (u-h)^2 + (h+F)^2|A(h,R)|\right\}|A(h,R)|^\varepsilon \\ &\le C\left\{\frac{1}{(R-r)^2} + \frac{(h+F)^2}{(h-k)^2}\right\}\frac{1}{(h-k)^{2\varepsilon}}\left(\int_{A(k,R)} (u-k)^2\right)^{1+\varepsilon}\end{aligned}$$

or

$$\|(u-h)^+\|_{L^2(B_r)} \le C\left\{\frac{1}{R-r} + \frac{h+F}{h-k}\right\}\frac{1}{(h-k)^\varepsilon}\|(u-k)^+\|^{1+\varepsilon}_{L^2(B_R)}. \tag{4.3}$$

Now we carry out the iteration. Set $\varphi(k,r) = \|(u-k)^+\|_{L^2(B_r)}$. For $\tau = \frac{1}{2}$ and some $k > 0$ to be determined. Define for $\ell = 0, 1, \dots,$

$$\begin{aligned} k_\ell &= k_0 + k\left(1 - \frac{1}{2^\ell}\right) \quad (\le k_0 + k), \\ r_\ell &= \tau + \frac{1}{2^\ell}(1-\tau). \end{aligned}$$

Obviously we have

$$k_\ell - k_{\ell-1} = \frac{k}{2^\ell}, \qquad r_{\ell-1} - r_\ell = \frac{1}{2^\ell}(1-\tau).$$

Therefore we have for $\ell = 0, 1, \ldots$

$$\varphi(k_\ell, r_\ell) \le C\left\{\frac{2^\ell}{1-\tau} + \frac{2^\ell(k_0+F+k)}{k}\right\}\frac{2^{\varepsilon\ell}}{k^\varepsilon}[\varphi(k_{\ell-1}, r_{\ell-1})]^{1+\varepsilon}$$
$$\le \frac{C}{1-\tau}\cdot\frac{k_0+F+k}{k^{1+\varepsilon}}\cdot 2^{(1+\varepsilon)\ell}\cdot[\varphi(k_{\ell-1}, r_{\ell-1})]^{1+\varepsilon}.$$

Next we prove inductively for any $\ell = 0, 1, \ldots,$

$$\varphi(k_\ell, r_\ell) \le \frac{\varphi(k_0, r_0)}{\gamma^\ell} \quad \text{for some } \gamma > 1 \tag{4.4}$$

if k is sufficiently large. Obviously it is true for $\ell = 0$. Suppose it is true for $\ell - 1$. We write

$$[\varphi(k_{\ell-1}, r_{\ell-1})]^{1+\varepsilon} \le \left\{\frac{\varphi(k_0, r_0)}{\gamma^{\ell-1}}\right\}^{1+\varepsilon} = \frac{\varphi(k_0, r_0)^\varepsilon}{\gamma^{\ell\varepsilon-(1+\varepsilon)}}\cdot\frac{\varphi(k_0, r_0)}{\gamma^\ell}.$$

Then we obtain

$$\varphi(k_\ell, r_\ell) \le \frac{C\gamma^{1+\varepsilon}}{1-\tau}\cdot\frac{k_0+F+k}{k^{1+\varepsilon}}\cdot[\varphi(k_0, r_0)]^\varepsilon\cdot\frac{2^{\ell(1+\varepsilon)}}{\gamma^{\ell\varepsilon}}\cdot\frac{\varphi(k_0, r_0)}{\gamma^\ell}.$$

Choose γ first such that $\gamma^\varepsilon = 2^{1+\varepsilon}$. Note $\gamma > 1$. Next, we need

$$\frac{C\gamma^{1+\varepsilon}}{1-\tau}\cdot\left(\frac{\varphi(k_0, r_0)}{k}\right)^\varepsilon\cdot\frac{k_0+F+k}{k} \le 1.$$

Therefore we choose

$$k = C_*\{k_0 + F + \varphi(k_0, r_0)\}$$

for C_* large. Let $\ell \to +\infty$ in (4.4). We conclude

$$\varphi(k_0 + k, \tau) = 0.$$

Hence we have

$$\sup_{B_{1/2}} u^+ \le (C_* + 1)\{k_0 + F + \varphi(k_0, r_0)\}.$$

Recall $k_0 = C\|u^+\|_{L^2(B_1)}$ and $\varphi(k_0, r_0) \le \|u^+\|_{L^2(B_1)}$. This finishes the proof. □

Next we give the second proof of Theorem 4.1.

METHOD 2: MOSER'S APPROACH: First we explain the idea. By choosing the test function appropriately, we will estimate the L^{p_1}-norm of u in a smaller ball by the L^{p_2}-norm of u for $p_1 > p_2$ in a larger ball, that is,

$$\|u\|_{L^{p_1}(B_{r_1})} \le C\|u\|_{L^{p_2}(B_{r_2})}$$

for $p_1 > p_2$ and $r_1 < r_2$. This is a reversed Hölder inequality. As a sacrifice C behaves like $\frac{1}{r_2-r_1}$. By iteration and a careful choice of $\{r_i\}$ and $\{p_i\}$, we will obtain the result.

For some $k > 0$ and $m > 0$, set $\overline{u} = u^+ + k$ and

$$\overline{u}_m = \begin{cases} \overline{u} & \text{if } u < m, \\ k + m & \text{if } u \geq m. \end{cases}$$

Then we have $D\overline{u}_m = 0$ in $\{u < 0\}$ and $\{u > m\}$ and $\overline{u}_m \leq \overline{u}$. Set the test function

$$\varphi = \eta^2\big(\overline{u}_m^\beta \overline{u} - k^{\beta+1}\big) \in H_0^1(B_1)$$

for some $\beta \geq 0$ and some nonnegative function $\eta \in C_0^1(B_1)$. Direct calculation yields

$$\begin{aligned} D\varphi &= \beta\eta^2\overline{u}_m^{\beta-1} D\overline{u}_m\overline{u} + \eta^2\overline{u}_m^\beta D\overline{u} + 2\eta D\eta\big(\overline{u}_m^\beta\overline{u} - k^{\beta+1}\big) \\ &= \eta^2\overline{u}_m^\beta(\beta D\overline{u}_m + D\overline{u}) + 2\eta D\eta\big(\overline{u}_m^\beta\overline{u} - k^{\beta+1}\big). \end{aligned}$$

We should emphasize that later on we will begin the iteration with $\beta = 0$. Note $\varphi = 0$ and $D\varphi = 0$ in $\{u \leq 0\}$. Hence if we substitute such φ in the equation we integrate in the set $\{u > 0\}$. Note also that $u^+ \leq \overline{u}$ and $\overline{u}_m^\beta\overline{u} - k^{\beta+1} \leq \overline{u}_m^\beta\overline{u}$ for $k > 0$.

First we have by the Hölder inequality

$$\begin{aligned} &\int a_{ij}D_iuD_j\varphi \\ &= \int a_{ij}D_i\overline{u}(\beta D_j\overline{u}_m + D_j\overline{u})\eta^2\overline{u}_m^\beta + 2\int a_{ij}D_i\overline{u}D_j\eta\big(\overline{u}_m^\beta\overline{u} - k^{\beta+1}\big)\eta \\ &\geq \lambda\beta\int \eta^2\overline{u}_m^\beta|D\overline{u}_m|^2 + \lambda\int\eta^2\overline{u}_m^\beta|D\overline{u}|^2 - \Lambda\int|D\overline{u}||D\eta|\overline{u}_m^\beta\overline{u}\eta \\ &\geq \lambda\beta\int \eta^2\overline{u}_m^\beta|D\overline{u}_m|^2 + \frac{\lambda}{2}\int\eta^2\overline{u}_m^\beta|D\overline{u}|^2 - \frac{2\Lambda^2}{\lambda}\int|D\eta|^2\overline{u}_m^\beta\overline{u}^2. \end{aligned}$$

Hence we obtain by noting $\overline{u} \geq k$

$$\begin{aligned} &\beta\int\eta^2\overline{u}_m^\beta|D\overline{u}_m|^2 + \int\eta^2\overline{u}_m^\beta|D\overline{u}|^2 \\ &\qquad \leq C\Big\{\int|D\eta|^2\overline{u}_m^\beta\overline{u}^2 + \int\big(|c|\eta^2\overline{u}_m^\beta\overline{u}^2 + |f|\eta^2\overline{u}_m^\beta\overline{u}\Big\} \\ &\qquad \leq C\Big\{\int|D\eta|^2\overline{u}_m^\beta\overline{u}^2 + \int c_0\eta^2\overline{u}_m^\beta\overline{u}^2\Big\}, \end{aligned}$$

where c_0 is defined as

$$c_0 = |c| + \frac{|f|}{k}.$$

Choose $k = \|f\|_{L^q}$ if f is not identically 0. Otherwise choose arbitrary $k > 0$ and eventually let $k \to 0^+$. By assumption we have

$$\|c_0\|_{L^q} \leq \Lambda + 1.$$

Set $w = \overline{u}_m^{\beta/2}\overline{u}$. Note

$$|Dw|^2 \leq (1+\beta)\big\{\beta\overline{u}_m^\beta|D\overline{u}_m|^2 + \overline{u}_m^\beta|D\overline{u}|^2\big\}.$$

Therefore we have

$$\int |Dw|^2\eta^2 \le C\left\{(1+\beta)\int w^2|D\eta|^2 + (1+\beta)\int c_0 w^2\eta^2\right\}$$

or

$$\int |D(w\eta)|^2 \le C\left\{(1+\beta)\int w^2|D\eta|^2 + (1+\beta)\int c_0 w^2\eta^2\right\}.$$

The Hölder inequality implies

$$\int c_0 w^2\eta^2 \le \left(\int c_0^q\right)^{\frac{1}{q}}\left(\int (\eta w)^{\frac{2q}{q-1}}\right)^{1-\frac{1}{q}} \le (\Lambda+1)\left(\int (\eta w)^{\frac{2q}{q-1}}\right)^{1-\frac{1}{q}}.$$

By interpolation inequality and the Sobolev inequality with $2^* = \frac{2n}{n-2} > \frac{2q}{q-1} > 2$ if $q > \frac{n}{2}$, we have

$$\begin{aligned}\|\eta w\|_{L^{\frac{2q}{q-1}}} &\le \varepsilon\|\eta w\|_{L^{2^*}} + C(n,q)\varepsilon^{-\frac{n}{2q-n}}\|\eta w\|_{L^2}\\ &\le \varepsilon\|D(\eta w)\|_{L^2} + C(n,q)\varepsilon^{-\frac{n}{2q-n}}\|\eta w\|_{L^2}\end{aligned}$$

for any small $\varepsilon > 0$. Therefore we obtain

$$\int |D(w\eta)|^2 \le C\left\{(1+\beta)\int w^2|D\eta|^2 + (1+\beta)^{\frac{2q}{2q-n}}\int w^2\eta^2\right\}$$

and in particular

$$\int |D(w\eta)|^2 \le C(1+\beta)^\alpha \int (|D\eta|^2+\eta^2)w^2,$$

where α is a positive number depending only on n and q. The Sobolev inequality then implies

$$\left(\int |\eta w|^{2\chi}\right)^{1/\chi} \le C(1+\beta)^\alpha \int (|D\eta|^2+\eta^2)w^2$$

where $\chi = \frac{n}{n-2} > 1$ for $n > 2$ and $\chi > 2$ for $n = 2$.

Choose the cutoff function as follows. For any $0 < r < R \le 1$ set $\eta \in C_0^1(B_R)$ with the property

$$\eta \equiv 1 \text{ in } B_r \quad \text{and} \quad |D\eta| \le \frac{2}{R-r}.$$

Then we obtain

$$\left(\int_{B_r} w^{2\chi}\right)^{1/\chi} \le C\frac{(1+\beta)^\alpha}{(R-r)^2}\int_{B_R} w^2.$$

Recalling the definition of w, we have

$$\left(\int_{B_r} \bar{u}^{2\chi}\bar{u}_m^{\beta\chi}\right)^{1/\chi} \le C\frac{(1+\beta)^\alpha}{(R-r)^2}\int_{B_R} \bar{u}^2\bar{u}_m^\beta.$$

Set $\gamma = \beta + 2 \geq 2$. Then we obtain

$$\left(\int_{B_r} \bar{u}_m^{\gamma\chi}\right)^{1/\chi} \leq C\frac{(\gamma-1)^\alpha}{(R-r)^2}\int_{B_R} \bar{u}^\gamma$$

provided the integral in the right-hand side is bounded. By letting $m \to +\infty$ we conclude that

$$\|\bar{u}\|_{L^{\gamma\chi}(B_r)} \leq \left(C\frac{(\gamma-1)^\alpha}{(R-r)^2}\right)^{1/\gamma} \|\bar{u}\|_{L^\gamma(B_R)}$$

provided $\|\bar{u}\|_{L^\gamma(B_R)} < +\infty$, where $C = C(n, q, \lambda, \Lambda)$ is a positive constant independent of γ. The above estimate suggests that we iterate, beginning with $\gamma = 2$, as $2, 2\chi, 2\chi^2, \ldots$. Now set for $i = 0, 1, \ldots,$

$$\gamma_i = 2\chi^i \quad \text{and} \quad r_i = \frac{1}{2} + \frac{1}{2^{i+1}}.$$

By $\gamma_i = \chi\gamma_{i-1}$ and $r_{i-1} - r_i = 1/2^{i+1}$, we have for $i = 1, 2, \ldots,$

$$\|\bar{u}\|_{L^{\gamma_i}(B_{r_i})} \leq C(n, q, \lambda, \Lambda)^{\frac{i}{\chi^i}} \|\bar{u}\|_{L^{\gamma_{i-1}}(B_{r_{i-1}})}$$

provided $\|\bar{u}\|_{L^{\gamma_{i-1}}(B_{r_{i-1}})} < +\infty$. Hence by iteration we obtain

$$\|\bar{u}\|_{L^{\gamma_i}(B_{r_i})} \leq C^{\sum \frac{i}{\chi^i}} \|\bar{u}\|_{L^2(B_1)};$$

in particular,

$$\left(\int_{B_{1/2}} \bar{u}^{2\chi^i}\right)^{\frac{1}{2\chi^i}} \leq C\left(\int_{B_1} \bar{u}^2\right)^{\frac{1}{2}}.$$

Letting $i \to +\infty$ we get

$$\sup_{B_{1/2}} \bar{u} \leq C\|\bar{u}\|_{L^2(B_1)} \quad \text{or} \quad \sup_{B_{1/2}} u^+ \leq C\{\|u^+\|_{L^2(B_1)} + k\}.$$

Recall the definition of k. This finishes the proof for $p = 2$. □

REMARK 4.2. If the subsolution u is bounded, we may simply take the test function

$$\varphi = \eta^2(\bar{u}^{\beta+1} - k^{\beta+1}) \in H_0^1(B_1)$$

for some $\beta \geq 0$ and some nonnegative function $\eta \in C_0^1(B_1)$.

Next we discuss the general p case of Theorem 4.1. This is based on a dilation argument.

Take any $R \leq 1$. Define

$$\tilde{u}(y) = u(Ry) \quad \text{for } y \in B_1.$$

It is easy to see that $\tilde{u}$ satisfies the following equation:

$$\int_{B_1} \tilde{a}_{ij} D_i\tilde{u} D_j\varphi + \tilde{c}\tilde{u}\varphi \leq \int_{B_1} \tilde{f}\varphi \quad \text{for any } \varphi \in H_0^1(B_1) \text{ and } \phi \geq 0 \text{ in } B_1$$

where

$$\tilde{a}(y) = a(Ry), \quad \tilde{c}(y) = R^2 c(Ry), \quad \tilde{f}(y) = R^2 f(Ry),$$

for any $y \in B_1$. Direct calculation shows

$$|\tilde{a}_{ij}|_{L^\infty(B_1)} + \|\tilde{c}\|_{L^q(B_1)} = |a_{ij}|_{L^\infty(B_R)} + R^{2-\frac{n}{q}}\|c\|_{L^q(B_R)} \leq \Lambda.$$

We may apply what we just proved to $\tilde{u}$ in B_1 and rewrite the result in terms of u. Hence we obtain for $p \geq 2$

$$\sup_{B_{R/2}} u^+ \leq C\left\{\frac{1}{R^{n/p}}\|u^+\|_{L^p(B_R)} + R^{2-\frac{n}{q}}\|f\|_{L^q(B_R)}\right\}$$

where $C = C(n, \lambda, \Lambda, p, q)$ is a positive constant. The estimate in $B_{\theta R}$ can be obtained by applying the above result to $B_{(1-\theta)R}(y)$ for any $y \in B_{\theta R}$. Take $R = 1$. This is Theorem 4.1 for any $\theta \in (0, 1)$ and $p \geq 2$.

Now we prove the statement for $p \in (0, 2)$. We showed that for any $\theta \in (0, 1)$ and $0 < R \leq 1$ there holds

$$\begin{aligned}\|u^+\|_{L^\infty(B_{\theta R})} &\leq C\left\{\frac{1}{[(1-\theta)R]^{n/2}}\|u^+\|_{L^2(B_R)} + R^{2-\frac{n}{q}}\|f\|_{L^q(B_R)}\right\} \\ &\leq C\left\{\frac{1}{[(1-\theta)R]^{n/2}}\|u^+\|_{L^2(B_R)} + \|f\|_{L^q(B_1)}\right\}.\end{aligned}$$

For $p \in (0, 2)$ we have

$$\int_{B_R} (u^+)^2 \leq \|u^+\|^{2-p}_{L^\infty(B_R)} \int_{B_R} (u^+)^p$$

and hence by the Hölder inequality

$$\begin{aligned}&\|u^+\|_{L^\infty(B_{\theta R})} \\ &\leq C\left\{\frac{1}{[(1-\theta)R]^{n/2}}\|u^+\|^{1-p/2}_{L^\infty(B_R)}\left(\int_{B_R} (u^+)^p\,dx\right)^{\frac{1}{2}} + \|f\|_{L^q(B_R)}\right\} \\ &\leq \frac{1}{2}\|u^+\|_{L^\infty(B_R)} + C\left\{\frac{1}{[(1-\theta)R]^{n/p}}\left(\int_{B_R} (u^+)^p\right)^{\frac{1}{p}} + \|f\|_{L^q(B_R)}\right\}.\end{aligned}$$

Set $f(t) = \|u^+\|_{L^\infty(B_t)}$ for $t \in (0, 1]$. Then for any $0 < r < R \leq 1$

$$f(r) \leq \frac{1}{2} f(R) + \frac{C}{(R-r)^{n/p}}\|u^+\|_{L^p(B_1)} + C\|f\|_{L^q(B_1)}.$$

We apply the following lemma to get for any $0 < r < R < 1$

$$f(r) \leq \frac{C}{(R-r)^{\frac{n}{p}}}\|u^+\|_{L^p(B_1)} + C\|f\|_{L^q(B_1)}.$$

Let $R \to 1^-$. We obtain for any $\theta < 1$

$$\|u^+\|_{L^\infty(B_\theta)} \leq \frac{C}{(1-\theta)^{\frac{n}{p}}}\|u^+\|_{L^p(B_1)} + C\|f\|_{L^q(B_1)}.$$

□

We need the following simple lemma:

LEMMA 4.3 *Let $f(t) \geq 0$ be bounded in $[\tau_0, \tau_1]$ with $\tau_0 \geq 0$. Suppose for $\tau_0 \leq t < s \leq \tau_1$ we have*

$$f(t) \leq \theta f(s) + \frac{A}{(s-t)^\alpha} + B$$

for some $\theta \in [0,1)$. Then for any $\tau_0 \leq t < s \leq \tau_1$ there holds

$$f(t) \leq c(\alpha, \theta)\left\{\frac{A}{(s-t)^\alpha} + B\right\}.$$

PROOF: Fix $\tau_0 \leq t < s \leq \tau_1$. For some $0 < \tau < 1$ we consider the sequence $\{t_i\}$ defined by

$$t_0 = t \quad \text{and} \quad t_{i+1} = t_i + (1-\tau)\tau^i(s-t).$$

Note $t_\infty = s$. By iteration

$$f(t) = f(t_0) \leq \theta^k f(t_k) + \left[\frac{A}{(1-\tau)^\alpha}(s-t)^{-\alpha} + B\right]\sum_{i=0}^{k-1}\theta^i\tau^{-i\alpha}.$$

Choose $\tau < 1$ such that $\theta\tau^{-\alpha} < 1$, that is, $\theta < \tau^\alpha < 1$. As $k \to \infty$ we have

$$f(t) \leq c(\alpha, \theta)\left\{\frac{A}{(1-\tau)^\alpha}(s-t)^{-\alpha} + B\right\}.$$

□

In the rest of this section we use Moser's iteration to prove a high integrability result that is closely related to Theorem 4.1. For the next result we require $n \geq 3$.

THEOREM 4.4 *Suppose $a_{ij} \in L^\infty(B_1)$ and $c \in L^{n/2}(B_1)$ satisfy the following assumption:*

$$\lambda|\xi|^2 \leq a_{ij}(x)\xi_i\xi_j \leq \Lambda|\xi|^2 \quad \textit{for any } x \in B_1,\ \xi \in \mathbb{R}^n,$$

for some positive constants λ and Λ. Suppose that $u \in H^1(B_1)$ is a subsolution in the following sense:

$$\int_{B_1} a_{ij}D_iuD_j\varphi + cu\varphi \leq \int_{B_1} f\varphi \quad \textit{for any } \varphi \in H_0^1(B_1) \textit{ and } \varphi \geq 0 \textit{ in } B_1.$$

If $f \in L^q(B_1)$ for some $q \in [\frac{2n}{n+2}, \frac{n}{2})$, then $u^+ \in L^{q^*}_{\text{loc}}(B_1)$ for $\frac{1}{q^*} = \frac{1}{q} - \frac{2}{n}$. Moreover, there holds

$$\|u^+\|_{L^{q^*}(B_{1/2})} \leq C\{\|u^+\|_{L^2(B_1)} + \|f\|_{L^q(B_1)}\}$$

where $C = C(n, \lambda, \Lambda, q, \varepsilon(K))$ is a positive constant with

$$\varepsilon(K) = \left(\int_{\{|c|>K\}} |c|^{n/2}\right)^{2/n}.$$

PROOF: For $m > 0$, set $\overline{u} = u^+$ and

$$\overline{u}_m = \begin{cases} \overline{u} & \text{if } u < m, \\ m & \text{if } u \geq m. \end{cases}$$

Then set the test function

$$\varphi = \eta^2 \overline{u}_m^\beta \overline{u} \in H_0^1(B_1)$$

for some $\beta \geq 0$ and some nonnegative function $\eta \in C_0^1(B_1)$. By similar calculations as in the proof of Theorem 4.1 we conclude

$$\left(\int \eta^{2\chi} \overline{u}_m^{\beta\chi} \overline{u}^{2\chi} \right)^{1/\chi} \leq$$

$$C(1+\beta) \left\{ \int |D\eta|^2 \overline{u}_m^\beta \overline{u}^2 + \int |c| \eta^2 \overline{u}_m^\beta \overline{u}^2 + \int |f| \eta^2 \overline{u}_m^\beta \overline{u} \right\}$$

where $\chi = \frac{n}{n-2} > 1$. The Hölder inequality implies for any $K > 0$

$$\begin{aligned} \int |c| \eta^2 \overline{u}_m^\beta \overline{u}^2 &\leq K \int_{\{|c| \leq K\}} \eta^2 \overline{u}_m^\beta \overline{u}^2 + \int_{\{|c| > K\}} |c| \eta^2 \overline{u}_m^\beta \overline{u}^2 \\ &\leq K \int \eta^2 \overline{u}_m^\beta \overline{u}^2 + \left(\int_{\{|c|>K\}} |c|^{\frac{n}{2}} \right)^{\frac{2}{n}} \left(\int (\eta^2 \overline{u}_m^\beta \overline{u}^2)^{\frac{n}{n-2}} \right)^{\frac{n-2}{n}} \\ &\leq K \int \eta^2 \overline{u}_m^\beta \overline{u}^2 + \varepsilon(K) \left(\int \eta^{2\chi} \overline{u}_m^{\beta\chi} \overline{u}^{2\chi} \right)^{\frac{1}{\chi}}. \end{aligned}$$

Note $\varepsilon(K) \to 0$ as $K \to +\infty$ since $c \in L^{n/2}(B_1)$. Hence for bounded β we obtain by choosing large $K = K(\beta)$

$$\left(\int \eta^{2\chi} \overline{u}_m^{\beta\chi} \overline{u}^{2\chi} \right)^{1/\chi} \leq C(1+\beta) \left\{ \int (|D\eta|^2 + \eta^2) \overline{u}_m^\beta \overline{u}^2 + \int |f| \eta^2 \overline{u}_m^\beta \overline{u} \right\}.$$

Observe

$$\overline{u}_m^\beta \overline{u} \leq \overline{u}_m^{\beta - \frac{\beta}{\beta+2}} \overline{u}^{1 + \frac{\beta}{\beta+2}} = \left(\overline{u}_m^\beta \overline{u}^2 \right)^{\frac{\beta+1}{\beta+2}}.$$

Therefore by the Hölder inequality again we have for $\eta \leq 1$

$$\begin{aligned} \int |f| \eta^2 \overline{u}_m^\beta \overline{u} &\leq \left(\int |f|^q \right)^{\frac{1}{q}} \left(\int (\eta^2 \overline{u}_m^\beta \overline{u}^2)^\chi \right)^{\frac{\beta+1}{(\beta+2)\chi}} |\mathrm{supp}\, \eta|^{1 - \frac{1}{q} - \frac{\beta+1}{(\beta+2)\chi}} \\ &\leq \varepsilon \left(\int \eta^{2\chi} \overline{u}^\chi \overline{u}_m^{\beta\chi} \right)^{\frac{1}{\chi}} + C(\varepsilon, \beta) \left(\int |f|^q \right)^{\frac{\beta+2}{q}}, \end{aligned}$$

provided

$$1 - \frac{1}{q} - \frac{\beta+1}{(\beta+2)\chi} \geq 0 \quad \text{which is equivalent to} \quad \beta + 2 \leq \frac{q(n-2)}{n-2q}.$$

Hence β is required to be bounded, depending only on n and q. Then we obtain

$$\left(\int \eta^{2\chi}\bar{u}_m^{\beta\chi}\bar{u}^{2\chi}\right)^{1/\chi} \leq C\left\{\int (|D\eta|^2+\eta^2)\bar{u}_m^{\beta}\bar{u}^2 + \|f\|_{L^q}^{\beta+2}\right\}.$$

By setting $\gamma = \beta + 2$, we have by definition of q^*

$$2 \leq \gamma \leq \frac{q(n-2)}{n-2q} = \frac{q^*}{\chi}. \tag{4.5}$$

We conclude, as before, for any such γ in (4.5) and any $0 < r < R \leq 1$

$$\|\bar{u}\|_{L^{\chi\gamma}(B_r)} \leq C\left\{\frac{1}{(R-r)^{2/\gamma}}\|\bar{u}\|_{L^\gamma(B_R)} + \|f\|_{L^q(B_1)}\right\} \tag{4.6}$$

provided $\|\bar{u}\|_{L^\gamma(B_R)} < +\infty$. Again this suggests the iteration $2, 2\chi, 2\chi^2, \ldots$.

For given $q \in [\frac{2n}{n+2}, \frac{n}{2})$, there exists a positive integer k such that

$$2\chi^{k-1} \leq \frac{q(n-2)}{n-2q} < 2\chi^k.$$

Hence for such k we get by finitely many iterations of (4.6)

$$\|\bar{u}\|_{L^{2\chi^k}(B_{3/4})} \leq C\{\|\bar{u}\|_{L^2(B_1)} + \|f\|_{L^q(B_1)}\};$$

in particular,

$$\|\bar{u}\|_{L^{\frac{q^*}{\chi}}(B_{3/4})} \leq C\{\|\bar{u}\|_{L^2(B_1)} + \|f\|_{L^q(B_1)}\},$$

while with $\gamma = q^*/\chi$ in (4.6) we obtain

$$\|\bar{u}\|_{L^{q^*}(B_{1/2})} \leq C\{\|\bar{u}\|_{L^{q^*/\chi}(B_{3/4})} + \|f\|_{L^q(B_1)}\}.$$

This finishes the proof. □

4.3. Hölder Continuity

We first discuss homogeneous equations with no lower-order terms. Consider

$$Lu \equiv -D_i(a_{ij}(x)D_ju) \quad \text{in } B_1(0) \subset \mathbb{R}^n$$

where $a_{ij} \in L^\infty(B_1)$ satisfies

$$\lambda|\xi|^2 \leq a_{ij}(x)\xi_i\xi_j \leq \Lambda|\xi|^2 \quad \text{for all } x \in B_1(0) \text{ and } \xi \in \mathbb{R}^n$$

for some positive constants λ and Λ.

DEFINITION 4.5 The function $u \in H^1_{\text{loc}}(B_1)$ is called a *subsolution* (*supersolution*) of the equation $Lu = 0$ if

$$\int_{B_1} a_{ij}D_iuD_j\varphi \leq 0 \ \ (\geq 0) \quad \text{for all } \varphi \in H^1_0(B_1) \text{ and } \varphi \geq 0.$$

LEMMA 4.6 *Let $\Phi \in C^{0,1}_{\text{loc}}(\mathbb{R})$ be convex. Then:*

(i) *If u is a subsolution and $\Phi' \geq 0$, then $v = \Phi(u)$ is also a subsolution provided $v \in H^1_{\text{loc}}(B_1)$.*

(ii) *If u is a supersolution and $\Phi' \leq 0$, then $v = \Phi(u)$ is a subsolution provided $v \in H^1_{\text{loc}}(B_1)$.*

REMARK 4.7. If u is a subsolution, then $(u-k)^+$ is also a subsolution, where $(u-k)^+ = \max\{0, u-k\}$. In this case $\Phi(s) = (s-k)^+$.

PROOF: We prove by direct computation.

(i) Assume first $\Phi \in C^2_{\text{loc}}(\mathbb{R})$. Then

$$\Phi'(s) \geq 0, \quad \Phi''(s) \geq 0.$$

Consider $\varphi \in C^1_0(B_1)$ with $\varphi \geq 0$. Direct calculation yields

$$\begin{aligned}\int_{B_1} a_{ij} D_i v D_j \varphi &= \int_{B_1} a_{ij} \Phi'(u) D_i u D_j \varphi \\ &= \int_{B_1} a_{ij} D_i u D_j(\Phi'(u)\varphi) - \int_{B_1} (a_{ij} D_i u D_j u)\varphi\Phi''(u) \leq 0,\end{aligned}$$

where $\Phi'(u)\varphi \in H^1_0(B_1)$ is nonnegative. In general, set $\Phi_\varepsilon(s) = \rho_\varepsilon * \Phi(s)$ with ρ_ε as the standard mollifier. Then $\Phi'_\varepsilon(s) = \rho_\varepsilon * \Phi'(s) \geq 0$ and $\Phi''_\varepsilon(s) \geq 0$. Hence $\Phi_\varepsilon(u)$ is a subsolution by what we just proved. Note $\Phi'_\varepsilon(s) \to \Phi'(s)$ a.e. as $\varepsilon \to 0^+$. Hence the Lebesgue dominant covergence theorem implies the result.

(ii) This is proved similarly. □

We need the following Poincaré-Sobolev inequality:

LEMMA 4.8 *For any $\varepsilon > 0$ there exists a $C = C(\varepsilon, n)$ such that for $u \in H^1(B_1)$ with*

$$|\{x \in B_1; u = 0\}| \geq \varepsilon |B_1| \quad \textit{there holds} \quad \int_{B_1} u^2 \leq C \int_{B_1} |Du|^2.$$

PROOF: Suppose not. Then there exists a sequence $\{u_m\} \subset H^1(B_1)$ such that

$$|\{x \in B_1; u_m = 0\}| \geq \varepsilon |B_1|, \quad \int_{B_1} u_m^2 = 1, \quad \int_{B_1} |Du_m|^2 \to 0 \text{ as } m \to \infty.$$

Hence we may assume $u_m \to u_0 \in H^1(B_1)$ strongly in $L^2(B_1)$ and weakly in $H^1(B_1)$. Clearly u_0 is a nonzero constant. So

$$\begin{aligned}0 = \lim_{m\to\infty} \int_{B_1} |u_m - u_0|^2 &\geq \lim_{m\to\infty} \int_{\{u_m=0\}} |u_m - u_0|^2 \\ &\geq |u_0|^2 \inf_m |\{u_m = 0\}| > 0.\end{aligned}$$

Contradiction. □

THEOREM 4.9 (Density Theorem) *Suppose u is a positive supersolution in B_2 with*

$$|\{x \in B_1 : u \geq 1\}| \geq \varepsilon |B_1|.$$

Then there exists a constant C depending only on ε, n, and Λ/λ such that

$$\inf_{B_{1/2}} u \geq C.$$

PROOF: We may assume that $u \geq \delta > 0$. Then let $\delta \to 0+$. By Lemma 4.6, $v = (\log u)^-$ is a subsolution, bounded by $\log \delta^{-1}$. Then Theorem 4.1 yields

$$\sup_{B_{1/2}} v \leq C \left(\int_{B_1} |v|^2 \right)^{1/2}.$$

Note $|\{x \in B_1 : v = 0\}| = |\{x \in B_1 : u \geq 1\}| \geq \varepsilon |B_1|$. Lemma 4.8 implies

$$\sup_{B_{1/2}} v \leq C \left(\int_{B_1} |Dv|^2 \right)^{1/2}. \tag{4.7}$$

We will prove that the right-hand side is bounded. To this end, set $\varphi = \frac{\zeta^2}{u}$ for $\zeta \in C_0^1(B_2)$ as the text function. Then we obtain

$$0 \leq \int a_{ij} D_i u D_j \left(\frac{\zeta^2}{u} \right) = - \int \zeta^2 \frac{a_{ij} D_i u D_j u}{u^2} + 2 \int \frac{\zeta a_{ij} D_i u D_j \zeta}{u},$$

which implies

$$\int \zeta^2 |D \log u|^2 \leq C \int |D\zeta|^2.$$

So for fixed $\zeta \in C_0^1(B_2)$ with $\zeta \equiv 1$ in B_1 we have

$$\int_{B_1} |D \log u|^2 \leq C.$$

Combining this with (4.7) we obtain

$$\sup_{B_{1/2}} v = \sup_{B_{1/2}} (\log u)^- \leq C \quad \text{which gives} \quad \inf_{B_{1/2}} u \geq e^{-C} > 0.$$

□

THEOREM 4.10 (Oscillation Theorem) *Suppose that u is a bounded solution of $Lu = 0$ in B_2. Then there exists a $\gamma = \gamma(n, \frac{\Lambda}{\lambda}) \in (0, 1)$ such that*

$$\operatorname{osc}_{B_{1/2}} u \leq \gamma \operatorname{osc}_{B_1} u.$$

PROOF: In fact, local boundedness is proved in the previous section. Set

$$\alpha_1 = \sup_{B_1} u \quad \text{and} \quad \beta_1 = \inf_{B_1} u.$$

Consider the solution

$$\frac{u - \beta_1}{\alpha_1 - \beta_1} \quad \text{or} \quad \frac{\alpha_1 - u}{\alpha_1 - \beta_1}.$$

Note the following equivalence:

$$u \geq \frac{1}{2}(\alpha_1 + \beta_1) \Longleftrightarrow \frac{u - \beta_1}{\alpha_1 - \beta_1} \geq \frac{1}{2},$$
$$u \leq \frac{1}{2}(\alpha_1 + \beta_1) \Longleftrightarrow \frac{\alpha_1 - u}{\alpha_1 - \beta_1} \geq \frac{1}{2}.$$

CASE 1. Suppose that

$$\left|\left\{x \in B_1 : \frac{2(u - \beta_1)}{\alpha_1 - \beta_1} \geq 1\right\}\right| \geq \frac{1}{2}|B_1|.$$

Apply the above theorem to $\frac{u-\beta_1}{\alpha_1-\beta_1} \geq 0$ in B_1. We have for some $C > 1$

$$\inf_{B_{1/2}} \frac{u - \beta_1}{\alpha_1 - \beta_1} \geq \frac{1}{C},$$

which results in the following estimate:

$$\inf_{B_{1/2}} u \geq \beta_1 + \frac{1}{C}(\alpha_1 - \beta_1).$$

CASE 2. Suppose

$$\left|\left\{x \in B_1 : \frac{2(\alpha_1 - u)}{\alpha_1 - \beta_1} \geq 1\right\}\right| \geq \frac{1}{2}|B_1|.$$

Similarly as in Case 1 we obtain

$$\sup_{B_{1/2}} u \leq \alpha_1 - \frac{1}{C}(\alpha_1 - \beta_1).$$

Now set

$$\alpha_2 = \sup_{B_{1/2}} u \quad \text{and} \quad \beta_2 = \inf_{B_{1/2}} u.$$

Note $\beta_2 \geq \beta_1$ and $\alpha_2 \leq \alpha_1$. In both cases, we have

$$\alpha_2 - \beta_2 \leq \left(1 - \frac{1}{C}\right)(\alpha_1 - \beta_1).$$

□

The De Giorgi theorem is an easy consequence of the above results.

THEOREM 4.11 (De Giorgi) *Suppose $Lu = 0$ weakly in B_1. Then there holds*

$$\sup_{B_{1/2}} |u(x)| + \sup_{x,y \in B_{1/2}} \frac{|u(x) - u(y)|}{|x - y|^\alpha} \leq c\left(n, \frac{\Lambda}{\lambda}\right)\|u\|_{L^2(B_1)}$$

with $\alpha = \alpha(n, \frac{\Lambda}{\lambda}) \in (0, 1)$.

In the rest of the section we will discuss the Hölder continuity of solutions to general linear equations. We need the following lemma:

LEMMA 4.12 *Suppose that* $a_{ij} \in L^\infty(B_r)$ *satisfies*

$$\lambda|\xi|^2 \le a_{ij}(x)\xi_i\xi_j \le \Lambda|\xi|^2 \quad \textit{for any } x \in B_r,\ \xi \in \mathbb{R}^n,$$

for some $0 < \lambda \le \Lambda < +\infty$. *Suppose* $u \in H^1(B_r)$ *satisfies*

$$\int_{B_r} a_{ij} D_i u D_j \varphi = 0 \quad \textit{for any } \varphi \in H_0^1(B_r).$$

Then there exists an $\alpha \in (0,1)$ *such that for any* $\rho < r$ *there holds*

$$\int_{B_\rho} |Du|^2 \le C\left(\frac{\rho}{r}\right)^{n-2+2\alpha} \int_{B_r} |Du|^2$$

where C *and* α *depend only on* n *and* $\frac{\Lambda}{\lambda}$.

PROOF: By dilation, consider $r = 1$. We restrict our consideration to the range $\rho \in (0, \frac{1}{4}]$, since it is trivial for $\rho \in (\frac{1}{4}, 1]$. We may further assume that $\int_{B_1} u = 0$ since the function $u - |B_1|^{-1}\int_{B_1} u$ solves the same equation. The Poincaré inequality yields

$$\int_{B_1} u^2 \le c(n) \int_{B_1} |Du|^2.$$

Hence Theorem 4.11 implies for $|x| \le \frac{1}{2}$

$$|u(x) - u(0)|^2 \le C|x|^{2\alpha} \int_{B_1} |Du|^2$$

where $\alpha \in (0,1)$ is determined in Theorem 4.11. For any $0 < \rho \le \frac{1}{4}$ take a cutoff function $\zeta \in C_0^\infty(B_{2\rho})$ with $\zeta \equiv 1$ in B_ρ and $0 \le \rho \le 1$ and $|D\zeta| \le \frac{2}{\rho}$. Then set $\varphi = \zeta^2(u - u(0))$. Hence the equation yields

$$\begin{aligned} 0 &= \int_{B_1} a_{ij} D_i u\big(\zeta^2 D_j u + 2\zeta D_j\zeta(u - u(0))\big) \\ &\ge \frac{\lambda}{2} \int_{B_{2\rho}} \zeta^2 |Du|^2 - C \sup_{B_{2\rho}} |u - u(0)|^2 \int_{B_{2\rho}} |D\zeta|^2. \end{aligned}$$

Therefore we have

$$\int_{B_\rho} |Du|^2 \le C\rho^{n-2} \sup_{B_{2\rho}} |u - u(0)|^2.$$

The conclusion follows easily. □

Now we may prove the following result in the same way we proved Theorem 3.8, with Lemma 3.10 replaced by Lemma 4.12.

THEOREM 4.13 *Assume* $a_{ij} \in L^\infty(B_1)$ *and* $c \in L^n(B_1)$ *satisfies*

$$\lambda|\xi|^2 \le a_{ij}(x)\xi_i\xi_j \le \Lambda|\xi|^2 \quad \text{for any } x \in B_1,\ \xi \in \mathbb{R}^n,$$

for some $0 < \lambda \le \Lambda < +\infty$. *Suppose that* $u \in H^1(B_1)$ *satisfies*

$$\int_{B_1} a_{ij} D_j u D_i \varphi + cu\varphi = \int_{B_1} f\varphi \quad \text{for any } \varphi \in H_0^1(B_1).$$

If $f \in L^q(B_1)$ *for some* $q > \frac{n}{2}$, *then* $u \in C^\alpha(B_1)$ *for some* $\alpha = \alpha(n, q, \lambda, \Lambda, \|c\|_{L^n}) \in (0,1)$. *Moreover, there exists* $R_0 = R_0(q, \lambda, \Lambda, \|c\|_{L^n})$ *such that for any* $x \in B_{1/2}$ *and* $r \le R_0$ *there holds*

$$\int_{B_r(x)} |Du|^2 \le Cr^{n-2+2\alpha}\{\|f\|^2_{L^q(B_1)} + \|u\|^2_{H^1(B_1)}\}$$

where $C = C(n, q, \lambda, \Lambda, \|c\|_{L^n})$ *is a positive constant.*

4.4. Moser's Harnack Inequality

In this section we only discuss equations without lower-order terms. Suppose Ω is a domain in $\mathbb{R}^n$. We always assume that $a_{ij} \in L^\infty(\Omega)$ satisfies

$$\lambda|\xi|^2 \le a_{ij}(x)\xi_i\xi_j \le \Lambda|\xi|^2 \quad \text{for all } x \in \Omega \text{ and } \xi \in \mathbb{R}^n$$

for some positive constants λ and Λ.

THEOREM 4.14 (Local Boundedness) *Let* $u \in H^1(\Omega)$ *be a nonnegative subsolution in* Ω *in the following sense*:

$$\int_\Omega a_{ij} D_i u D_j \varphi \le \int_\Omega f\varphi \quad \text{for any } \varphi \in H_0^1(\Omega) \text{ and } \phi \ge 0 \text{ in } \Omega.$$

Suppose $f \in L^q(\Omega)$ *for some* $q > \frac{n}{2}$. *Then there holds for any* $B_R \subset \Omega$, *any* $0 < r < R$, *and any* $p > 0$

$$\sup_{B_r} u \le C\left\{\frac{1}{(R-r)^{n/p}}\|u^+\|_{L^p(B_R)} + R^{2-\frac{n}{q}}\|f\|_{L^q(B_R)}\right\}$$

where $C = C(n, \lambda, \Lambda, p, q)$ *is a positive constant.*

PROOF: This is a special case of Theorem 4.1 in the dilated version. □

THEOREM 4.15 (Weak Harnack Inequality) *Let* $u \in H^1(\Omega)$ *be a nonnegative supersolution in* Ω *in the following sense*:

$$(*) \qquad \int_\Omega a_{ij} D_i u D_j \varphi \ge \int_\Omega f\varphi \quad \text{for any } \varphi \in H_0^1(\Omega) \text{ and } \varphi \ge 0 \text{ in } \Omega.$$

Suppose $f \in L^q(\Omega)$ for some $q > \frac{n}{2}$. Then for any $B_R \subset \Omega$ there holds for any $0 < p < \frac{n}{n-2}$ and any $0 < \theta < \tau < 1$

$$\inf_{B_{\theta R}} u + R^{2-\frac{n}{q}} \|f\|_{L^q(B_R)} \geq C\left(\frac{1}{R^n}\int_{B_{\tau R}} u^p\right)^{\frac{1}{p}}$$

where C depends only on n, p, q, λ, Λ, θ, and τ.

PROOF: We prove for $R = 1$.

Step 1. We prove that the result holds for some $p_0 > 0$.

Set $\bar{u} = u + k > 0$ for some $k > 0$ to be determined and $v = \bar{u}^{-1}$. First we will derive the equation for v. For any $\varphi \in H_0^1(B_1)$ with $\varphi \geq 0$ in B_1 consider $\bar{u}^{-2}\varphi$ as the test function in $(*)$. We have

$$\int_{B_1} a_{ij} D_i u \frac{D_j\varphi}{\bar{u}^2} - 2\int_{B_1} a_{ij} D_i u D_j \bar{u} \frac{\varphi}{\bar{u}^3} \geq \int_{B_1} f \frac{\varphi}{\bar{u}^2}.$$

Note $D\bar{u} = Du$ and $Dv = -\bar{u}^2 D\bar{u}$. Therefore we obtain

$$\int_{B_1} a_{ij} D_j v D_i \varphi + \tilde{f} v\varphi \leq 0 \quad \text{where we set} \quad \tilde{f} = \frac{f}{\bar{u}}.$$

In other words, v is a nonnegative subsolution to some homogeneous equation. Choose $k = \|f\|_{L^q}$ if f is not identically 0. Otherwise choose arbitrary $k > 0$ and then let $k \to 0^+$. Note

$$\|\tilde{f}\|_{L^q(B_1)} \leq 1.$$

Thus Theorem 4.1 implies that for any $\tau \in (\theta, 1)$ and any $p > 0$

$$\sup_{B_\theta} \bar{u}^{-p} \leq C\int_{B_\tau} \bar{u}^{-p},$$

that is,

$$\inf_{B_\theta} \bar{u} \geq C\left(\int_{B_\tau} \bar{u}^{-p}\,dx\right)^{-\frac{1}{p}} = C\left(\int_{B_\tau} \bar{u}^{-p}\int_{B_\tau} \bar{u}^p\right)^{-\frac{1}{p}}\left(\int_{B_\tau} \bar{u}^p\right)^{\frac{1}{p}}$$

where $C = C(n, q, p, \lambda, \Lambda, \tau, \theta) > 0$.

The key point is to show that there exists a $p_0 > 0$ such that

$$\int_{B_\tau} \bar{u}^{-p_0} \cdot \int_{B_\tau} \bar{u}^{p_0} \leq C(n, q, \lambda, \Lambda, \tau).$$

We will show that for any $\tau < 1$ there holds

$$\int_{B_\tau} e^{p_0|w|} \leq C(n, q, \lambda, \Lambda, \tau) \tag{4.8}$$

where $w = \log\bar{u} - \beta$ with $\beta = |B_\tau|^{-1}\int_{B_\tau} \log\bar{u}$.

We have two methods:

(1) Prove directly.
(2) Prove that $w \in \text{BMO}$, that is, for any $B_r(y) \subset B_1(0)$

$$\frac{1}{r^n}\int_{B_r} |w - w_{y,r}|dx \le C.$$

Then (4.8) follows from Theorem 3.5 (John-Nirenberg lemma).

We shall prove (4.8) directly first. Recall $\overline{u} = u + k \ge k > 0$. Note that

$$e^{p_0|w|} = 1 + p_0|w| + \frac{(p_0|w|)^2}{2!} + \cdots + \frac{(p_0|w|)^n}{n!} + \cdots.$$

Hence we need to estimate $\int_{B_\tau} |w|^\beta$ for each positive integer β.

We first derive the equation for w. Consider $\overline{u}^{-1}\varphi$ as test function in $(*)$. Here we need $\varphi \in L^\infty(B_1) \cap H_0^1(B_1)$ with $\varphi \ge 0$. By direct calculation as before and by the fact that $Dw = \overline{u}^{-1}D\overline{u}$, we have

$$\int_{B_1} a_{ij}D_iwD_jw\varphi \le \int_{B_1} a_{ij}D_iwD_j\varphi + \int_{B_1}(-\tilde{f}\varphi) \tag{4.9}$$

for any $\varphi \in L^\infty(B_1) \cap H_0^1(B_1)$ with $\varphi \ge 0$. Replace φ by φ^2 in (4.9). The Hölder inequality implies

$$\int_{B_1} |Dw|^2\varphi^2 \le C\left\{\int_{B_1} |D\varphi|^2 + \int_{B_1} |\tilde{f}|\varphi^2\right\}.$$

By the Hölder and Sobolev inequalities we obtain

$$\int_{B_1} |\tilde{f}|\varphi^2 \le \|\tilde{f}\|_{L^{n/2}}\|\varphi\|^2_{L^{2n/(n-2)}} \le c(n,q)\|D\varphi\|^2_{L^2}.$$

Therefore we have

$$\int_{B_1} |Dw|^2\varphi^2 \le C\int_{B_1} |D\varphi|^2 \tag{4.10}$$

with $C = C(n,q,\lambda,\Lambda) > 0$. Take $\varphi \in C_0^1(B_1)$ with $\varphi \equiv 1$ in B_τ. Then we obtain

$$\int_{B_\tau} |Dw|^2 \le C(n,q,\lambda,\Lambda,\tau). \tag{4.11}$$

Hence the Poincaré inequality implies

$$\int_{B_\tau} w^2 \le c(n,\tau)\int_{B_\tau} |Dw|^2 \le C(n,q,\lambda,\Lambda,\tau)$$

since $\int_{B_\tau} w = 0$. Furthermore, we conclude from (4.10)

$$\int_{B_{\tau'}} w^2 \le C(n,q,\lambda,\Lambda,\tau,\tau') \tag{4.12}$$

for any $\tau' \in (\tau, 1)$.

Next we will estimate $\int_{B_\tau} |w|^\beta$ for any $\beta \geq 2$. Choose $\varphi = \zeta^2 |w_m|^{2\beta} \in H_0^1(B_1) \cap L^\infty(B_1)$ with

$$w_m = \begin{cases} -m, & w \leq -m, \\ w, & |w| < m, \\ m, & w \geq m. \end{cases}$$

Substitute such φ in (4.9) to get

$$\begin{aligned} \int_{B_1} \zeta^2 |w_m|^{2\beta} a_{ij} D_i w D_j w \leq (2\beta) \int_{B_1} \zeta^2 a_{ij} D_i w D_j |w_m| |w_m|^{2\beta-1} \\ + \int_{B_1} 2\zeta |w_m|^{2\beta} a_{ij} D_i w D_j \zeta + \int_{B_1} |\widetilde{f}| \zeta^2 |w_m|^{2\beta}. \end{aligned}$$

Note $a_{ij} D_i w D_j |w_m| = a_{ij} D_i w_m D_j |w_m| \leq a_{ij} D_i w_m D_j w_m$ a.e. in B_1. Young's inequality implies

$$\begin{aligned} (2\beta)|w_m|^{2\beta-1} &\leq \frac{2\beta - 1}{2\beta} |w_m|^{2\beta} + \frac{1}{2\beta}(2\beta)^{2\beta} \\ &= \left(1 - \frac{1}{2\beta}\right) |w_m|^{2\beta} + (2\beta)^{2\beta-1}. \end{aligned}$$

Hence we obtain

$$\begin{aligned} \int_{B_1} \zeta^2 |w_m|^{2\beta} a_{ij} D_i w D_j w \leq \left(1 - \frac{1}{2\beta}\right) \int_{B_1} \zeta^2 |w_m|^{2\beta} a_{ij} D_i w_m D_j w_m \\ + (2\beta)^{2\beta-1} \int_{B_1} \zeta^2 a_{ij} D_i w_m D_j w_m \\ + \int_{B_1} 2\zeta |w_m|^{2\beta} a_{ij} D_i w D_j \zeta + \int_{B_1} |\widetilde{f}| \zeta^2 |w_m|^{2\beta} \end{aligned}$$

and hence

$$\begin{aligned} &\int_{B_1} \zeta^2 |w_m|^{2\beta} a_{ij} D_i w D_j w \\ &\quad \leq (2\beta)^{2\beta} \int_{B_1} \zeta^2 a_{ij} D_i w_m D_j w_m \\ &\quad\quad + (4\beta) \int_{B_1} \zeta |w_m|^{2\beta} a_{ij} D_i w D_j \zeta + 2\beta \int_{B_1} |\widetilde{f}| \zeta^2 |w_m|^{2\beta}. \end{aligned}$$

Therefore we obtain

$$\int_{B_1} \zeta^2 |w_m|^{2\beta} |Dw|^2 \le C\Big\{(2\beta)^{2\beta} \int_{B_1} \zeta^2 |Dw_m|^2 + \beta \int_{B_1} \zeta |w_m|^{2\beta} |Dw||D\zeta| + \beta \int_{B_1} |\tilde{f}|\zeta^2 |w_m|^{2\beta}\Big\}.$$

Note that the first term in the right side is bounded in (4.11). Applying the Cauchy inequality to the second term in the right side we conclude

$$\int_{B_1} \zeta^2 |w_m|^{2\beta} |Dw|^2 \le C\Big\{(2\beta)^{2\beta} \int_{B_1} \zeta^2 |Dw_m|^2 + \beta^2 \int_{B_1} |w_m|^{2\beta} |D\zeta|^2 + \beta \int_{B_1} |\tilde{f}|\zeta^2 |w_m|^{2\beta}\Big\}.$$

Note $Dw = Dw_m$ for $|w| < m$ and $Dw_m = 0$ for $|w| > m$. Hence we have

$$\int_{B_1} \zeta^2 |w_m|^{2\beta} |Dw_m|^2 \le C\Big\{(2\beta)^{2\beta} \int_{B_1} \zeta^2 |Dw_m|^2 + \beta^2 \int_{B_1} |w_m|^{2\beta} |D\zeta|^2 + \beta \int_{B_1} |\tilde{f}|\zeta^2 |w_m|^{2\beta}\Big\}.$$

In the following, we write $w = w_m$ and then let $m \to +\infty$. By Young's inequality we obtain

$$\begin{aligned} |D(\zeta|w|^\beta)|^2 &\le 2|D\zeta|^2|w|^{2\beta} + 2\beta^2\zeta^2|w|^{2\beta-2}|Dw|^2 \\ &\le 2|D\zeta|^2|w|^{2\beta} + 2\zeta^2|Dw|^2\Big(\frac{\beta-1}{\beta}|w|^{2\beta} + \frac{1}{\beta}\beta^{2\beta}\Big) \end{aligned}$$

and hence

$$\begin{aligned} \int_{B_1} |D(\zeta|w|^\beta)|^2 \le C\Big\{(2\beta)^{2\beta} &\int_{B_1} \zeta^2|Dw|^2 \\ &+ \beta^2 \int |D\zeta|^2|w|^{2\beta} + \beta \int_{B_1} |\tilde{f}|\zeta^2|w|^{2\beta}\Big\}. \end{aligned}$$

The Hölder inequality implies

$$\int_{B_1} |\tilde{f}|\zeta^2|w|^{2\beta} \le \Big(\int_{B_1} |\tilde{f}|^q\Big)^{\frac{1}{q}} \Big(\int_{B_1} (\zeta|w|^\beta)^{\frac{2q}{q-1}}\Big)^{1-\frac{1}{q}}.$$

By the interpolation and Sobolev inequalities with $2^* = \frac{2n}{n-2} > \frac{2q}{q-1} > 2$ if $q > \frac{n}{2}$, we have

$$\begin{aligned} \|\zeta|w|^\beta\|_{L^{\frac{2q}{q-1}}} &\le \varepsilon\|\zeta|w|^\beta\|_{L^{2^*}} + C(n,q)\varepsilon^{-\frac{n}{2q-n}}\|\zeta|w|^\beta\|_{L^2} \\ &\le \varepsilon\|D(\zeta|w|^\beta)\|_{L^2} + C(n,q)\varepsilon^{-\frac{n}{2q-n}}\|\zeta|w|^\beta\|_{L^2} \end{aligned}$$

for any small $\varepsilon > 0$. Therefore we obtain by (4.10)

$$\int_{B_1} |D(\zeta|w|^\beta)|^2 \leq C\left\{(2\beta)^{2\beta}\int_{B_1}\zeta^2|Dw|^2 + \beta^\alpha\int_{B_1}(|D\zeta|^2+\zeta^2)|w|^{2\beta}\right\}$$

$$\leq C\left\{(2\beta)^{2\beta}\int_{B_1}|D\zeta|^2 + \beta^\alpha\int_{B_1}(|D\zeta|^2+\zeta^2)|w|^{2\beta}\right\}$$

for some positive constant α depending only on n and q. Apply the Sobolev inequality for $\zeta|w|^\beta \in W_0^{1,2}(\mathbb{R}^n)$ with $\chi = \frac{n}{n-2}$ to get

$$\left(\int_{B_1}\zeta^{2\chi}|w|^{2\beta\chi}\right)^{1/\chi} \leq C\beta^\alpha\left\{(2\beta)^{2\beta}\int_{B_1}|D\zeta|^2 + \int_{B_1}(|D\zeta|^2+\zeta^2)|w|^{2\beta}\right\}.$$

Choose the cutoff function as follows: For $\tau \leq r < R \leq 1$, set $\zeta \equiv 1$ on $B_r(0)$, $\zeta \equiv 0$ in $B_1(0)\setminus B_R(0)$, and $|D\zeta| \leq \frac{2}{R-r}$. Therefore we have

$$\left(\int_{B_r}|w|^{2\beta\chi}\right)^{1/\chi} \leq \frac{C\beta^\alpha}{(R-r)^2}\left\{(2\beta)^{2\beta} + \int_{B_R}|w|^{2\beta}\right\}.$$

For some $\tau' \in (\tau, 1)$ set $\beta_i = \chi^{i-1}$ and $r_i = \tau + \frac{1}{2^{i-1}}(\tau'-\tau)$ for any $i = 1, 2, \ldots$. Then for each $i = 1, 2, \ldots,$

$$\left(\int_{B_{r_i}}|w|^{2\chi^i}\right)^{1/\chi} \leq \frac{C\chi^{(i-1)\alpha}2^{2(i-1)}}{(\tau'-\tau)^2}\left\{(2\chi^{i-1})^{2\chi^{i-1}} + \int_{B_{r_{i-1}}}|w|^{2\chi^{i-1}}\right\}.$$

Set

$$I_j = \|w\|_{L^{2\chi^j}(B_{r_j})}.$$

Then we have for $j = 1, 2, \ldots,$

$$I_j \leq C^{\frac{j}{2\chi^j}}\{2\chi^{j-1} + I_{j-1}\}$$

with $C = C(n, q, \lambda, \Lambda, \tau, \tau') > 0$. Iterating the above inequality and observing that

$$\sum_{i=0}^{\infty}\frac{i}{\chi^i} < \infty,$$

we obtain

$$I_j \leq C\sum_{i=1}^{j}\chi^{i-1} + CI_0, \quad \text{that is,} \quad I_j \leq C\chi^j + CI_0.$$

Now for $\beta \geq 2$ there exists a j such that $2\chi^{j-1} \leq \beta < 2\chi^j$. Hence

$$I_\beta(B_\tau) \equiv \left(\int_{B_\tau}|w|^\beta\right)^{1/\beta} \leq CI_j \leq C\chi^j + CI_0 \leq C\beta + CI_0 \leq C_0\beta,$$

since I_0 is bounded in (4.12). Hence we obtain for $\beta \geq 1$

$$\int_{B_\tau} |w|^\beta dx \leq C_0^\beta \beta^\beta \leq C_0^\beta e^\beta \beta!$$

where we used the Sterling formula for integer β. Hence for integer $\beta \geq 1$

$$\int_{B_\tau} \frac{(p_0|w|)^\beta}{\beta!} \leq p_0^\beta (C_0 e)^\beta \leq \frac{1}{2^\beta}$$

by choosing $p_0 = (2C_0 e)^{-1}$. This proves that

$$\int e^{p_0|w|} = \int 1 + p_0|w| + \frac{(p_0|w|)^2}{2!} + \cdots \leq 1 + \frac{1}{2^1} + \frac{1}{2^2} + \cdots \leq 2.$$

REMARK 4.16. The above method, avoiding BMO, is elementary in nature.

Now we give the second proof of estimate (4.8). Estimate (4.10) gives

$$\int_{B_1} |Dw|^2 \zeta^2 \leq C \int_{B_1} |D\zeta|^2 \quad \text{for any } \zeta \in C_0^1(B_1).$$

Then for any $B_{2r}(y) \subset B_1$ choose ζ with

$$\operatorname{supp} \zeta \subset B_{2r}(y), \quad \zeta \equiv 1 \text{ in } B_r(y), \quad |D\zeta| \leq \frac{2}{r}.$$

Then we obtain

$$\int_{B_r(y)} |Dw|^2 \leq C r^{n-2}.$$

Hence the Poincaré inequality implies

$$\begin{aligned} \frac{1}{r^n} \int_{B_r(y)} |w - w_{y,r}| &\leq \frac{1}{r^{n/2}} \Big(\int_{B_r(y)} |w - w_{y,r}|^2 \Big)^{1/2} \\ &\leq \frac{1}{r^{n/2}} \Big(r^2 \int_{B_r(y)} |Dw|^2 \Big)^{1/2} \leq C, \end{aligned}$$

that is, $w \in$ BMO. Then the John-Nirenberg lemma implies

$$\int_{B_\tau} e^{p_0|w|} \leq C.$$

Step 2. The result holds for any positive $p < \frac{n}{n-2}$.

We need to prove for any $0 < r_1 < r_2 < 1$ and $0 < p_2 < p_1 < \frac{n}{n-2}$ there holds

$$\Big(\int_{B_{r_1}} \bar{u}^{p_1} \Big)^{\frac{1}{p_1}} \leq C \Big(\int_{B_{r_2}} \bar{u}^{p_2} \Big)^{\frac{1}{p_2}} \tag{4.13}$$

for some $C = C(n, q, \lambda, \Lambda, r_1, r_2, p_1, p_2) > 0$. A similar calculation may be found in Section 4.2 (Method 2). Here we just point out some key steps.

Take $\varphi = \bar{u}^{-\beta}\eta^2$ for $\beta \in (0,1)$ as the test function in $(*)$. Then we have

$$\int_{B_1} |D\bar{u}|^2 \bar{u}^{-\beta-1}\eta^2 \leq C\left\{\frac{1}{\beta^2}\int_{B_1} |D\eta|^2 \bar{u}^{1-\beta} + \frac{1}{\beta}\int_{B_1} \frac{|f|}{k}\eta^2\bar{u}^{1-\beta}\right\}.$$

Set $\gamma = 1 - \beta \in (0,1)$ and $w = \bar{u}^{\gamma/2}$. Then we have

$$\int |Dw|^2\eta^2 \leq \frac{C}{(1-\gamma)^\alpha}\int w^2(|D\eta|^2 + \eta^2)$$

or

$$\int |D(w\eta)|^2 \leq \frac{C}{(1-\gamma)^\alpha}\int w^2(|D\eta|^2 + \eta^2)$$

for some positive $\alpha > 0$. The Sobolev embedding theorem and an appropriate choice for the cutoff function imply, with $\chi = \frac{n}{n-2}$, that for any $0 < r < R < 1$

$$\left(\int_{B_r} w^{2\chi}\right)^{\frac{1}{\chi}} \leq \frac{C}{(1-\gamma)^\alpha}\cdot\frac{1}{(R-r)^2}\int_{B_R} w^2$$

or

$$\left(\int_{B_r} \bar{u}^{\gamma\chi}\right)^{\frac{1}{\gamma\chi}} \leq \left(\frac{C}{(1-\gamma)^\alpha}\frac{1}{(R-r)^2}\right)^{\frac{1}{\gamma}}\left(\int_{B_R} \bar{u}^{\gamma}\right)^{\frac{1}{\gamma}}.$$

This holds for any $\gamma \in (0,1)$. Now (4.13) follows after finitely many iterations. □

Now the Harnack inequality is an easy consequence of the above results.

THEOREM 4.17 (Moser's Harnack Inequality) *Let $u \in H^1(\Omega)$ be a nonnegative solution in Ω*

$$\int_\Omega a_{ij}D_iuD_j\varphi = \int_\Omega f\varphi \quad \textit{for any } \varphi \in H_0^1(\Omega).$$

Suppose $f \in L^q(\Omega)$ for some $q > \frac{n}{2}$. Then there holds for any $B_R \subset \Omega$

$$\max_{B_R} u \leq C\{\min_{B_{R/2}} u + R^{2-\frac{n}{q}}\|f\|_{L^q(B_R)}\}$$

where $C = C(n, \lambda, \Lambda, q)$ is a positive constant.

COROLLARY 4.18 (Hölder Continuity) *Let $u \in H^1(\Omega)$ be a solution in Ω*

$$\int_\Omega a_{ij}D_iuD_j\varphi = \int_\Omega f\varphi \quad \textit{for any } \varphi \in H_0^1(\Omega).$$

Suppose $f \in L^q(\Omega)$ for some $q > \frac{n}{2}$. Then $u \in C^\alpha(\Omega)$ for some $\alpha \in (0,1)$ depending only on n, q, λ, and Λ. Moreover, there holds for any $B_R \subset \Omega$

$$|u(x) - u(y)| \leq C\left(\frac{|x-y|}{R}\right)^\alpha \left\{\left(\frac{1}{R^n}\int_{B_R} u^2\right)^{1/2} + R^{2-\frac{n}{q}}\|f\|_{L^q(B_R)}\right\}$$

for any $x, y \in B_{R/2}$ where $C = C(n, \lambda, \Lambda, q)$ is a positive constant.

PROOF: We prove the estimate for $R = 1$. Let $M(r) = \max_{B_r} u$ and $m(r) = \min_{B_r} u$ for $r \in (0,1)$. Then $M(r) < +\infty$ and $m(r) > -\infty$. It suffices to show that

$$\omega(r) \triangleq M(r) - m(r) \leq Cr^\alpha \left\{\left(\int_{B_1} u^2\right)^{1/2} + \|f\|_{L^q(B_1)}\right\} \quad \text{for any } r < \frac{1}{2}.$$

Set $\delta = 2 - \frac{n}{q}$. Apply Theorem 4.17 to $M(r) - u \geq 0$ in B_r to get

$$\sup_{B_{r/2}} (M(r) - u) \leq C\{\inf_{B_{r/2}} (M(r) - u) + r^\delta \|f\|_{L^q(B_r)}\},$$

that is,

$$M(r) - m\left(\frac{r}{2}\right) \leq C\left\{\left(M(r) - M\left(\frac{r}{2}\right)\right) + r^\delta \|f\|_{L^q(B_r)}\right\}. \tag{4.14}$$

Similarly, apply Harnack to $u - m(r) \geq 0$ in B_r to get

$$M\left(\frac{r}{2}\right) - m(r) \leq C\left\{\left(m\left(\frac{r}{2}\right) - m(r)\right) + r^\delta \|f\|_{L^q(B_r)}\right\}. \tag{4.15}$$

Then by adding (4.14) and (4.15) together we get

$$\omega(r) + \omega\left(\frac{r}{2}\right) \leq C\left\{\left(\omega(r) - \omega\left(\frac{r}{2}\right)\right) + r^\delta \|f\|_{L^q(B_r)}\right\}$$

or

$$\omega\left(\frac{r}{2}\right) \leq \gamma\omega(r) + Cr^\delta \|f\|_{L^q(B_r)}$$

for some $\gamma = \frac{C-1}{C+1} < 1$.

Apply Lemma 4.19 below with μ chosen such that $\alpha = (1-\mu)\log\gamma/\log\tau < \mu\delta$. We obtain

$$\omega(\rho) \leq C\rho^\alpha \left\{\omega\left(\frac{1}{2}\right) + \|f\|_{L^q(B_1)}\right\} \quad \text{for any } \rho \in (0, \tfrac{1}{2}].$$

While Theorem 4.14 implies

$$\omega\left(\frac{1}{2}\right) \leq C\left\{\left(\int_{B_1} u^2\right)^{1/2} + \|f\|_{L^q(B_1)}\right\}.$$

□

LEMMA 4.19 *Let ω and σ be nondecreasing functions in an interval $(0, R]$. Suppose there holds for all $r \le R$*

$$\omega(\tau r) \le \gamma\omega(r) + \sigma(r)$$

for some $0 < \gamma, \tau < 1$. Then for any $\mu \in (0,1)$ and $r \le R$ we have

$$\omega(r) \le C\left\{\left(\frac{r}{R}\right)^{\alpha}\omega(R) + \sigma(r^{\mu}R^{1-\mu})\right\}$$

where $C = C(\gamma, \tau)$ and $\alpha = \alpha(\gamma, \tau, \mu)$ are positive constants. In fact, $\alpha = (1-\mu)\log\gamma/\log\tau$.

PROOF: Fix some number $r_1 \le R$. Then for any $r \le r_1$ we have

$$\omega(\tau r) \le \gamma\omega(r) + \sigma(r_1)$$

since σ is nondecreasing. We now iterate this inequality to get for any positive integer k

$$\omega(\tau^k r_1) \le \gamma^k\omega(r_1) + \sigma(r_1)\sum_{i=0}^{k-1}\gamma^i \le \gamma^k\omega(R) + \frac{\sigma(r_1)}{1-\gamma}.$$

For any $r \le r_1$ we choose k in such a way that

$$\tau^k r_1 < r \le \tau^{k-1} r_1.$$

Hence we have

$$\omega(r) \le \omega(\tau^{k-1}r_1) \le \gamma^{k-1}\omega(R) + \frac{\sigma(r_1)}{1-\gamma} \le \frac{1}{\gamma}\left(\frac{r}{r_1}\right)^{\frac{\log\gamma}{\log\tau}}\omega(R) + \frac{\sigma(r_1)}{1-\gamma}.$$

Now let $r_1 = r^{\mu}R^{1-\mu}$. We obtain

$$\omega(r) \le \frac{1}{\gamma}\left(\frac{r}{R}\right)^{(1-\mu)\frac{\log\gamma}{\log\tau}}\omega(R) + \frac{\sigma(r^{\mu}R^{1-\mu})}{1-\gamma}.$$

This finishes the proof. □

COROLLARY 4.20 (Liouville Theorem) *Suppose u is a solution to a homogeneous equation in $\mathbb{R}^n$*

$$\int_{\mathbb{R}^n} a_{ij} D_i u D_j \varphi = 0 \quad \textit{for any } \varphi \in H_0^1(\mathbb{R}^n).$$

If u is bounded, then u is a constant.

PROOF: We showed that there exists a $\gamma < 1$ such that

$$\omega(r) \leq \gamma \omega(2r).$$

By iteration we have

$$\omega(r) \leq \gamma^k \omega(2^k r) \to 0 \quad \text{as } k \to \infty$$

since $\omega(2^k r) \leq C$ if u is bounded. Hence for any $r > 0$, $\omega(r) = 0$. □

4.5. Nonlinear Equations

Up to now, we have been discussing linear equations of the form

$$-D_j(a_{ij}(x) D_i u) = f(x) \quad \text{in } B_1.$$

It is natural to ask how they generalize to nonlinear equations. To answer this question, let us consider the equation for a solution v with the form

$$v(x) = \Phi(u(x))$$

for some smooth function $\Phi : \mathbb{R} \to \mathbb{R}$ with $\Phi' \neq 0$. Any estimates for u can be translated to those for v. To find the equation for v, we write

$$u = \Psi(v)$$

with $\Psi = \Phi^{-1}$. Then by setting $\eta = \Psi'(v)\xi$ for $\xi \in C_0^\infty(B_1)$ we have

$$\begin{aligned} \int a_{ij} D_i u D_j \xi &= \int a_{ij} \Psi'(v) D_i v D_j \xi \\ &= \int a_{ij} D_i v D_j \eta - \int \frac{\Psi''(v)}{\Psi'(v)} a_{ij} D_i v D_j v \eta. \end{aligned}$$

Therefore, if u is a solution

$$\int a_{ij} D_i u D_j \xi = \int f(x) \xi \quad \text{for any } \xi \in H_0^1(B_1),$$

then v satisfies

$$\int a_{ij} D_i v D_j \eta = \int \left(\frac{\Psi''(v)}{\Psi'(v)} a_{ij} D_i v D_j v + \frac{1}{\Psi'(v)} f \right) \eta \quad \text{for any } \eta \in C_0^\infty(B_1).$$

Note that the nonlinear term has quadratic growth in terms of Dv. Hence we may extend the space of test functions to $H_0^1(B_1) \cap L^\infty(B_1)$. It turns out that $H^1(B_1) \cap L^\infty(B_1)$ is also the right space for the solution. The following example illustrates that the boundedness of solutions is essential:

EXAMPLE. Consider the equation $-\Delta u = |Du|^2$ in the ball $B_R(0)$ in $\mathbb{R}^2$ with $R < 1$. It is easy to check that $u(x) = \log\log|x|^{-1} - \log\log R^{-1} \in H^1(B_R(0))$ is a weak solution with zero boundary data. Note that $u(x) \equiv 0$ is also a solution.

In this section we always assume $a_{ij} \in L^\infty(B_1)$ satisfies

$$\lambda|\xi|^2 \leq a_{ij}(x)\xi_i\xi_j \leq \Lambda|\xi|^2 \quad \text{for any } x \in B_1,\ \xi \in \mathbb{R}^n,$$

for some positive constants λ and Λ. We consider the nonlinear equation of the form

$$(*) \qquad \int a_{ij}(x) D_i u D_j \varphi = \int b(x, u, Du)\varphi \quad \text{for any } \varphi \in H_0^1(B_1) \cap L^\infty(B_1).$$

We say the nonlinear term b satisfies the natural growth condition if

$$|b(x,u,p)| \leq C(u)(f(x) + |p|^2) \quad \text{for any } (x,u,p) \in B_1 \times \mathbb{R} \times \mathbb{R}^n$$

for some constant $C(u)$ depending only on u and $f \in L^q(B_1)$ for some $q \geq \frac{2n}{n+2}$. We always assume

$$u \in H^1(B_1) \cap L^\infty(B_1).$$

LEMMA 4.21 *Suppose $u \in H^1(B_1)$ is a nonnegative solution of $(*)$ with $|u| \leq M$ in B_1 and that b satisfies the natural growth condition with $f(x) \in L^q(B_1)$ for some $q > \frac{n}{2}$. Then for any $B_R \subset B_1$ there holds*

$$\sup_{B_{R/2}} u \leq C\left\{\inf_{B_{R/2}} u + R^{2-\frac{n}{q}}\left(\int_{B_R} |f|^q\right)^{\frac{1}{q}}\right\}$$

where C is a positive constant depending only on n, λ, Λ, M, and q.

PROOF: Let $v = \frac{1}{\alpha}(e^{\alpha u} - 1)$ for some $\alpha > 0$. Then for $\varphi \in H_0^1(B_1) \cap L^\infty(B_1)$ with $\varphi \geq 0$ there holds

$$\begin{aligned}
\int a_{ij} D_i v D_j \varphi &= \int a_{ij} e^{\alpha u} D_i u D_j \varphi \\
&= \int a_{ij} D_i u D_j (e^{\alpha u}\varphi) - \alpha \int a_{ij} e^{\alpha u} D_i u D_j u \varphi \\
&= \int b(x,u,Du) e^{\alpha u}\varphi - \alpha \int a_{ij} e^{\alpha u} D_i u D_j u \varphi \\
&\leq C(M)\int (f(x) + |Du|^2) e^{\alpha u}\varphi - \alpha\lambda \int |Du|^2 e^{\alpha u}\varphi.
\end{aligned}$$

Hence by taking α large we have

$$(4.16) \qquad \begin{gathered} \int a_{ij} D_i v D_j \varphi \leq C \int f(x)\varphi \\ \text{for any } \varphi \in H_0^1(B_1) \cap L^\infty(B_1) \text{ with } \varphi \geq 0 \end{gathered}$$

for some positive constant C depending only on n, λ, Λ, and M. Observe that u and v are compatible. Therefore by Theorem 4.14 we obtain for any $p > 0$

$$\begin{aligned}\sup_{B_{R/2}} u \le C(M,\alpha) \sup_{B_{R/2}} v &\le C\left\{\left(\frac{1}{R^n}\int_{B_R} v^p\right)^{\frac{1}{p}} + R^{2-\frac{n}{q}}\left(\int_{B_R} f^q\right)^{\frac{1}{q}}\right\} \\ &\le C\left\{\left(\frac{1}{R^n}\int_{B_R} u^p\right)^{\frac{1}{p}} + R^{2-\frac{n}{q}}\left(\int_{B_R} f^q\right)^{\frac{1}{q}}\right\}.\end{aligned}$$

For the lower bound, we let $w = \frac{1}{\alpha}(1-e^{-\alpha u})$. As before, by choosing $\alpha > 0$ large we have

$$\int a_{ij} D_i w D_j \varphi \ge C \int f(x)\varphi \quad \text{for any } \varphi \in H_0^1(B_1) \cap L^\infty(B_1) \text{ with } \varphi \ge 0.$$

Hence by Theorem 4.15, we obtain for any $p \in (0, \frac{n}{n-2})$

$$\left(\frac{1}{R^n}\int_{B_R} u^p\right)^{\frac{1}{p}} \le C\left\{\inf_{B_{R/2}} u + R^{2-\frac{n}{q}}\left(\int_{B_R} f^q\right)^{\frac{1}{q}}\right\}.$$

Combining the above inequalities we prove Lemma 4.21. □

REMARK 4.22. In estimate (4.16) in the above proof, take $\varphi = (u + M)\eta^2$ for some $\eta \in C_0^1(B_1)$. Then by the Hölder inequality we conclude

$$\int |Du|^2\eta^2 \le C\left\{\int (|D\eta|^2 + |f|\eta^2)\right\}$$

for some positive constant C depending only on n, λ, Λ, and M. This implies the interior L^2-estimate of gradient Du in terms of these constants together with $\|f\|_{L^1(B_1)}$. This fact will be used in the proof of Theorem 4.24.

COROLLARY 4.23 *Suppose $u \in H^1(B_1)$ is a bounded solution of (∗) and that b satisfies the natural growth condition with $f(x) \in L^q(B_1)$ for some $q > \frac{n}{2}$. Then $u \in C^\alpha_{\text{loc}}(B_1)$ with $\alpha = \alpha(n, \lambda, \Lambda, q, |u|_{L^\infty})$. Moreover, there holds*

$$|u(x) - u(y)| \le C|x - y|^\alpha \quad \textit{for any } x, y \in B_{1/2}$$

where C is a positive constant depending only on n, λ, Λ, q, $|u|_{L^\infty(B_1)}$, and $\|f\|_{L^q(B_1)}$.

PROOF: The proof is identical to that of Corollary 4.18 with Theorem 4.17 replaced by Lemma 4.21. □

THEOREM 4.24 *Suppose $u \in H^1(B_1)$ is a bounded solution of (∗) and that b satisfies the natural growth condition with $f \in L^q(B_1)$ for some $q > n$. Assume further that $a_{ij} \in C^\alpha(B_1)$ for $\alpha = 1 - \frac{n}{q}$. Then $Du \in C^\alpha_{\text{loc}}(B_1)$. Moreover, there holds*

$$|Du|_{C^\alpha(B_{1/2})} \le C(n, \lambda, \Lambda, q, |u|_{L^\infty(B_1)}, \|f\|_{L^q(B_1)}).$$

PROOF: We only need to prove $Du \in L^\infty_{\text{loc}}$. Then the Hölder continuity is implied by Theorem 3.13. For any $B_r(x_0) \subset B_1$ solve for w such that

$$\int_{B_r(x_0)} a_{ij}(x_0) D_i w D_j \varphi = 0 \quad \text{for any } \varphi \in H_0^1(B_r(x_0))$$

with $w - u \in H_0^1(B_r(x_0))$. Then the maximum principle implies

$$\inf_{B_r(x_0)} u \le w \le \sup_{B_r(x_0)} u \quad \text{in } B_r(x_0)$$

or

$$\sup_{B_r(x_0)} |u - w| \le \operatorname{osc}_{B_r(x_0)} u. \tag{4.17}$$

By Lemma 3.10, we have for any $0 < \rho \le r$,

$$\int_{B_\rho(x_0)} |Du|^2 \le c\left\{\left(\frac{\rho}{r}\right)^n \int_{B_r(x_0)} |Du|^2 + \int_{B_r(x_0)} |D(u-w)|^2\right\} \tag{4.18}$$

and

$$\int_{B_\rho(x_0)} |Du - (Du)_{x_0,\rho}|^2 \le c\left\{\left(\frac{\rho}{r}\right)^{n+2} \int_{B_r(x_0)} |Du - (Du)_{x_0,r}|^2 + \int_{B_r(x_0)} |D(u-w)|^2\right\}. \tag{4.19}$$

Note that the function $v = u - w \in H_0^1(B_r(x_0))$ satisfies

$$\int_{B_r(x_0)} a_{ij}(x_0) D_i v D_j \varphi = \int_{B_r(x_0)} b(x,u,Du)\varphi + \int_{B_r(x_0)} \big(a_{ij}(x_0) - a_{ij}(x)\big) D_i u D_j \varphi,$$

$$\varphi \in H_0^1(B_r(x_0)) \cap L^\infty(B_r(x_0)).$$

Taking $\varphi = v$ and using the Sobolev inequality we obtain

$$\int_{B_r(x_0)} |Dv|^2 \le C\left\{\int_{B_r(x_0)} |Du|^2|v| + r^{2\alpha} \int_{B_r(x_0)} |Du|^2 + r^{n+2\alpha}\|f\|^2_{L^q(B_1)}\right\}.$$

Hence with (4.17) we conclude

$$\int_{B_r(x_0)} |Dv|^2 \le C\left\{\left(r^{2\alpha} + \operatorname{osc}_{B_r(x_0)} u\right) \int_{B_r(x_0)} |Du|^2 + r^{n+2\alpha}\|f\|^2_{L^q}\right\}. \tag{4.20}$$

Corollary 4.2 implies $u \in C^{\delta_0}$ for some $\delta_0 > 0$. Therefore we have by (4.18) and (4.20)

$$\int_{B_\rho(x_0)} |Du|^2 \le C\left\{\left[\left(\frac{\rho}{r}\right)^n + r^{2\alpha} + r^{\delta_0}\right] \int_{B_r(x_0)} |Du|^2 + r^{n+2\alpha}\|f\|^2_{L^q}\right\}.$$

By Lemma 3.4 we obtain that for any $\delta < 1$ there holds for any $B_r(x_0) \subset B_{7/8}$

$$\int_{B_r(x_0)} |Du|^2 \le Cr^{n-2+2\delta}\left\{\int_{B_{7/8}} |Du|^2 + \|f\|^2_{L^q(B_1)}\right\}.$$

This implies $u \in C^{\delta}_{\text{loc}}$ for any $\delta < 1$. Moreover, for any $B_r(x_0) \subset B_{3/4}$ there holds

$$\operatorname{osc}_{B_r(x_0)} u \le Cr^\delta$$

where C is some positive constant depending only on n, λ, Λ, q, $|u|_{L^\infty(B_1)}$, and $\|f\|_{L^q(B_1)}$, by Remark 4.22. With (4.20) we have for any $B_r(x_0) \subset B_{2/3}$

$$\begin{aligned}\int_{B_r(x_0)} |Dv|^2 &\le C\left\{(r^{2\alpha} + r^\delta)r^{n-2+2\delta} \int_{B_{7/8}} |Du|^2 + r^{n+2\alpha}\|f\|^2_{L^q}\right\}\\ &\le Cr^{n+2\alpha'}\end{aligned}$$

for some $\alpha' < \alpha$ if $\delta \in (0,1)$ is chosen such that $3\delta > 2$ and $\alpha + \delta > 1$. Hence with (4.19) we obtain for any $B_r(x_0) \subset B_{\frac{2}{3}}$ and any $0 < \rho \le r$

$$\int_{B_\rho(x_0)} |Du - (Du)_{x_0,\rho}|^2 \le C\left\{\left(\frac{\rho}{r}\right)^{n+2} \int_{B_r(x_0)} |Du - (Du)_{x_0,r}|^2 + r^{n+2\alpha'}\right\}.$$

By Lemma 3.4 and Theorem 3.1 we again conclude that $Du \in C^{\alpha'}_{\text{loc}}$ for some $\alpha' < \alpha$, in particular $Du \in L^\infty_{\text{loc}}$. This finishes the proof. $\square$

CHAPTER 5

Viscosity Solutions

5.1. Guide

In this chapter we generalize the notion of classical solutions to viscosity solutions and study their regularities. We define viscosity solutions by comparing them with quadratic polynomials and thus remove the requirement that solutions be at least C^2. The main tool for studying viscosity solutions is the maximum principle due to Alexandroff. We first generalize such maximum principles to viscosity solutions and then use the resulting estimate to discuss the regularity theory. We use it to control the distribution functions of solutions and obtain the Harnack inequality, and hence C^α regularity, which generalizes a result by Krylov and Safonov. We also use it to approximate solutions in L^∞ by quadratic polynomials and get Schauder ($C^{2,\alpha}$)-estimates. The methods are basically nonlinear in the sense that they do not rely on differentiating equations. This implies that the results obtained in this way may apply to general fully nonlinear equations, although in this chapter we focus only on linear equations.

Here we only try to explain a few basic ideas in obtaining estimates for viscosity solutions. Students should read the book **[4]** for further developments.

5.2. Alexandroff Maximum Principle

We begin this section with the definition of viscosity solutions. This very weak concept of solutions enables us to define a class of functions containing all classical solutions of linear and nonlinear elliptic equations with fixed ellipticity constants and bounded measurable coefficients.

Suppose that Ω is a bounded and connected domain in $\mathbb{R}^n$ and that $a_{ij} \in C(\Omega)$ satisfies

$$\lambda|\xi|^2 \le a_{ij}(x)\xi_i\xi_j \le \Lambda|\xi|^2 \quad \text{for any } x \in \Omega \text{ and any } \xi \in \mathbb{R}^n$$

for some positive constants λ and Λ. Consider the operator L in Ω defined by

$$Lu \equiv a_{ij}(x)D_{ij}u \quad \text{for } u \in C^2(\Omega).$$

Suppose $u \in C^2(\Omega)$ is a supersolution in Ω, that is, $Lu \le 0$. Then for any $\varphi \in C^2(\Omega)$ with $L\varphi > 0$ we have

$$L(u - \varphi) < 0 \quad \text{in } \Omega.$$

This implies by the maximum principle that $u - \varphi$ cannot have local interior minimums in Ω. In other words if $u - \varphi$ has a local minimum at $x_0 \in \Omega$, there holds

$$L\varphi(x_0) \le 0.$$

Geometrically $u - \varphi$ having a local minimum at x_0 means that φ touches u from below at x_0 if we adjust φ appropriately by adding a constant. This suggests the following definition. We assume $f \in C(\Omega)$.

DEFINITION 5.1 $u \in C(\Omega)$ is a viscosity supersolution (respectively, subsolution) of

$$Lu = f \quad \text{in } \Omega$$

if for any $x_0 \in \Omega$ and any function $\varphi \in C^2(\Omega)$ such that $u - \varphi$ has a local minimum (respectively, maximum) at x_0 there holds

$$L\varphi(x_0) \le f(x_0) \quad (\text{respectively, } L\varphi(x_0) \ge f(x_0)).$$

We say that u is a viscosity solution if it is a viscosity subsolution and a viscosity supersolution.

REMARK 5.2. By approximation we may replace the C^2-function φ by a quadratic polynomial Q.

REMARK 5.3. The above analysis shows that a classical supersolution is a viscosity supersolution. It is straightforward to prove that a C^2 viscosity supersolution is a classical supersolution. Similar statements hold for subsolutions and solutions.

REMARK 5.4. The notion of viscosity solutions can be generalized to nonlinear equations accordingly.

Now we define in a weak way the class of "all solutions to all elliptic equations." For any function φ that is C^2 at x_0, we have the following equivalence:

$$\begin{aligned}
&\sum_{i,j=1}^{n} a_{ij}(x_0) D_{ij}\varphi(x_0) \le 0 \\
&\qquad \Longleftrightarrow \sum_{k=1}^{n} \alpha_k e_k \le 0 \quad \text{with } \lambda \le \alpha_k \le \Lambda,\ e_k = e_k(D^2\varphi(x_0)) \\
&\qquad \Longleftrightarrow \sum_{e_i>0} \alpha_i e_i + \sum_{e_i<0} \alpha_i e_i \le 0 \\
&\qquad \Longleftrightarrow \sum_{e_i>0} \alpha_i e_i \le \sum_{e_i<0} \alpha_i(-e_i),
\end{aligned}$$

which implies

$$\lambda \sum_{e_i>0} e_i \le \Lambda \sum_{e_i<0} (-e_i)$$

where $e_1, \ldots, e_n$ are eigenvalues of the Hessian matrix $D^2\varphi(x_0)$. This means that positive eigenvalues of $D^2\varphi(x_0)$ are controlled by negative eigenvalues.

DEFINITION 5.5 Suppose f is a continuous function in Ω and that λ and Λ are two positive constants. We define $u \in C(\Omega)$ to belong to $\mathcal{S}^+(\lambda, \Lambda, f)$ (respectively,

$\mathcal{S}^-(\lambda, \Lambda, f)$) if for any $x_0 \in \Omega$ and any function $\varphi \in C^2(\Omega)$ such that $u - \varphi$ has a local minimum (respectively, maximum) at x_0 there holds

$$\lambda \sum_{e_i > 0} e_i(x_0) + \Lambda \sum_{e_i < 0} e_i(x_0) \le f(x_0)$$

$$\left(\text{respectively, } \Lambda \sum_{e_i > 0} e_i(x_0) + \lambda \sum_{e_i < 0} e_i(x_0) \ge f(x_0)\right)$$

where $e_1(x_0), \ldots, e_n(x_0)$ are eigenvalues of the Hessian matrix $D^2\varphi(x_0)$.

We denote $\mathcal{S}(\lambda, \Lambda, f) = \mathcal{S}^+(\lambda, \Lambda, f) \cap \mathcal{S}^-(\lambda, \Lambda, f)$.

REMARK 5.6. Any viscosity supersolutions of

$$a_{ij} D_{ij} u = f \quad \text{in } \Omega$$

belong to the class $\mathcal{S}^+(\lambda, \Lambda, f)$ where there holds

$$\lambda |\xi|^2 \le a_{ij}(x)\xi_i\xi_j \le \Lambda |\xi|^2 \quad \text{for any } x \in \Omega \text{ and any } \xi \in \mathbb{R}^n.$$

The class $\mathcal{S}^+(\lambda, \Lambda, f)$ and $\mathcal{S}^-(\lambda, \Lambda, f)$ also include solutions to fully nonlinear equations. Among them are the Pucci equations.

EXAMPLE. For any two positive constants $\lambda \le \Lambda$ let A be a symmetric matrix whose eigenvalues belong to $[\lambda, \Lambda]$, that is, $\lambda|\xi|^2 \le A_{ij}\xi_i\xi_j \le \Lambda|\xi|^2$ for any $\xi \in \mathbb{R}^n$. Let $\mathcal{A}_{\lambda,\Lambda}$ denote the class of all such matrices. For any symmetric matrix M we define the Pucci extremal operators

$$\mathcal{M}^-(M) = \mathcal{M}^-(\lambda, \Lambda, M) = \inf_{A \in \mathcal{A}_{\lambda,\Lambda}} A_{ij} M_{ij},$$

$$\mathcal{M}^+(M) = \mathcal{M}^+(\lambda, \Lambda, M) = \sup_{A \in \mathcal{A}_{\lambda,\Lambda}} A_{ij} M_{ij}.$$

Pucci's equations are given by

$$\mathcal{M}^-(\lambda, \Lambda, M) = f, \quad \mathcal{M}^+(\lambda, \Lambda, M) = g,$$

for continuous functions f and g in Ω. It is easy to see that

$$\mathcal{M}^-(\lambda, \Lambda, M) = \lambda \sum_{e_i > 0} e_i + \Lambda \sum_{e_i < 0} e_i,$$

$$\mathcal{M}^+(\lambda, \Lambda, M) = \Lambda \sum_{e_i > 0} e_i + \lambda \sum_{e_i < 0} e_i,$$

where $e_1, \ldots, e_n$ are eigenvalues of M. Therefore $u \in \mathcal{S}^+(\lambda, \Lambda, f)$ if and only if $\mathcal{M}^-(\lambda, \Lambda, D^2u) \le f$ in the viscosity sense; that is, for any $\varphi \in C^2(\Omega)$ such that $u - \varphi$ has a local minimum at $x_0 \in \Omega$ there holds

$$\mathcal{M}^-(\lambda, \Lambda, D^2\varphi(x_0)) \le f(x_0).$$

By the definition of $\mathcal{M}^-$ and $\mathcal{M}^+$ it is easy to check that for any two symmetric matrices M and N

$$\begin{aligned}\mathcal{M}^-(M) + \mathcal{M}^-(N) \le \mathcal{M}^-(M + N) \le \mathcal{M}^+(M) + \mathcal{M}^-(N) \\ \le \mathcal{M}^+(M + N) \le \mathcal{M}^+(M) + \mathcal{M}^+(N).\end{aligned}$$

This property will be needed in Section 5.4.

Next we derive the Alexandroff maximum principle for viscosity solutions. It replaces the energy inequality for solutions to equations of divergence forms. Let v be a continuous function in an open convex set Ω. Recall that the convex envelope of v in Ω is defined by

$$\Gamma(v)(x) = \sup_L \{L(x) : L \leq v \text{ in } \Omega, L \text{ an affine function}\}$$

for any $x \in \Omega$. It is easy to see that $\Gamma(v)$ is a convex function in Ω. The set $\{v = \Gamma(v)\} = \{x \in \Omega : v(x) = \Gamma(v)(x)\}$ is called the (lower) contact set of v. The points in the contact set are called contact points.

The following is the classical version of the Alexandroff maximum principle. We do not require that functions be solutions to elliptic equations. See Lemma 2.24.

LEMMA 5.7 *Suppose u is a $C^{1,1}$-function in B_1 with $u \geq 0$ on ∂B_1. Then there holds*

$$\sup_{B_1} u^- \leq c(n) \Big(\int_{B_1 \cap \{u = \Gamma_u\}} \det D^2 u \Big)^{\frac{1}{n}}$$

where Γ_u is the convex envelope of $-u^- = \min\{u, 0\}$.

Now we state the viscosity version.

THEOREM 5.8 *Suppose u belongs to $S^+(\lambda, \Lambda, f)$ in B_1 with $u \geq 0$ on ∂B_1 for some $f \in C(\Omega)$. Then there holds*

$$\sup_{B_1} u^- \leq c(n, \lambda, \Lambda) \Big(\int_{B_1 \cap \{u = \Gamma_u\}} (f^+)^n \Big)^{\frac{1}{n}}$$

where Γ_u is the convex envelope of $-u^- = \min\{u, 0\}$.

PROOF: We will prove that Γ_u is a $C^{1,1}$-function in B_1 and that at contact point x_0 there hold

$$f(x_0) \geq 0 \tag{5.1}$$

and

$$L(x) \leq \Gamma_u(x) \leq L(x) + C\{f(x_0) + \varepsilon(x)\}|x - x_0|^2 \tag{5.2}$$

for some affine function L and any x close to x_0, where $\varepsilon(x) \to 0$ as $x \to x_0$ and C is a positive constant depending only on n, λ, and Λ. We obtain by (5.2)

$$\det D^2 \Gamma_u(x) \leq C(n, \lambda, \Lambda)\big(f(x)\big)^n \quad \text{for a.e. } x \in \{u = \Gamma_u\}.$$

We may apply Lemma 5.7 to function Γ_u to get the result.

Suppose x_0 is a contact point, that is, $u(x_0) = \Gamma_u(x_0)$. We may assume $x_0 = 0$. We also assume, by subtracting a supporting plane at $x_0 = 0$, that $u \geq 0$ in B_1 and that $u(0) = 0$.

In order to prove (5.1) we take $h(x) = -\varepsilon|x|^2/2$ in B_1. Obviously $u - h$ has a minimum at 0. Note that the eigenvalues of $D^2h(0)$ are $-\varepsilon, \ldots, -\varepsilon$. By definition of $\mathcal{S}^+(\lambda, \Lambda, f)$ we have

$$-n\Lambda\varepsilon \le f(0).$$

By letting $\varepsilon \to 0$ we get (5.1).

For estimate (5.2) we will prove

$$0 \le \Gamma_u(x) \le C(n, \lambda, \Lambda)\{f(0) + \varepsilon(x)\}|x|^2 \quad \text{for } x \in B_1$$

where $\varepsilon(x) \to 0$ as $x \to 0$. By setting $w = \Gamma_u$ we need to estimate for any small $r > 0$

$$C_r = \frac{1}{r^2} \max_{B_r} w.$$

Fix $r > 0$. By convexity w attains its maximum in $\bar{B}_r$ at some point on the boundary, say, $(0, \ldots, 0, r)$. The set $\{x \in B_1 : w(x) \le w(0, \ldots, 0, r)\}$ is convex and contains B_r. It follows easily that

$$w(x', r) \ge w(0, \ldots, 0, r) = C_r r^2 \quad \text{for any } x = (x', r) \in B_1.$$

Take a positive number N to be determined. Set

$$R_r = \{(x', x_n) : |x'| \le Nr, \ |x_n| \le r\}.$$

We will construct a quadratic polynomial that touches u from below in R_r and curves upward very much. Set for some $b > 0$

$$h(x) = (x_n + r)^2 - b|x'|^2.$$

Then we have

(i) for $x_n = -r$, $h \le 0$;
(ii) for $|x'| = Nr$, $h \le (4 - bN^2)r^2 \le 0$ if we take $b = 4/N^2$;
(iii) for $x_n = r$, $h = 4r^2 - b|x'|^2 \le 4r^2$.

Hence if we set

$$\tilde{h}(x) = \frac{C_r}{4} h(x) = \frac{C_r}{4}\left\{(x_n + r)^2 - \frac{4}{N^2}|x'|^2\right\}$$

we obtain $\tilde{h} \le w \le u$ on ∂R_r (since w is the convex envelope of u) and $\tilde{h}(0) = C_r r^2/4 > 0 = w(0) = u(0)$. By lowering $\tilde{h}$ appropriately we conclude that $u - \tilde{h}$ has a local minimum somewhere inside R_r. Note the eigenvalues of $D^2\tilde{h}$ are given by $C_r/2, -2C_r/N^2, \ldots, -2C_r/N^2$. Hence by definition of $\mathcal{S}^+(\lambda, \Lambda, f)$ we have

$$\lambda \frac{C_r}{2} - 2\Lambda(n-1)\frac{C_r}{N^2} \le \max_{R_r} f.$$

By choosing N large, depending only on n, λ, and Λ, we obtain

$$C_r \le \frac{4}{\lambda} \max_{R_r} f \quad \text{or} \quad \max_{B_r} w \le \frac{4}{\lambda} r^2 \max_{R_r} f.$$

Note $\max_{R_r} f \to f(0)$ as $r \to 0$. This finishes the proof. □

We end this section with a simple consequence of Calderon-Zygmund decomposition. We first recall some terminology. Let Q_1 be the unit cube. Cut it equally into 2^n cubes, which we take as the first generation. Do the same cutting for these small cubes to get the second generation. Continue this process. These cubes (from all generations) are called *dyadic cubes*. Any $(k+1)$-generation cube Q comes from some k-generation cube $\widetilde{Q}$, which is called the *predecessor* of Q.

LEMMA 5.9 *Suppose measurable sets $A \subset B \subset Q_1$ have the following properties:*

(i) $|A| < \delta$ *for some* $\delta \in (0,1)$;
(ii) *for any dyadic cube Q, $|A \cap Q| \geq \delta|Q|$ implies $\widetilde{Q} \subset B$ for the predecessor $\widetilde{Q}$ of Q.*

Then there holds $|A| \leq \delta|B|$.

PROOF: Apply Calderon-Zygmund decomposition (Lemma 3.7) to $f = \chi_A$. We obtain, by assumption (i), a sequence of dyadic cubes $\{Q^j\}$ such that

$$A \subset \bigcup_j Q^j \quad \text{except for a set of measure } 0,$$

$$\delta \leq \frac{|A \cap Q^j|}{|Q^j|} < 2^n\delta, \qquad \frac{|A \cap \widetilde{Q}^j|}{|\widetilde{Q}^j|} < \delta,$$

for any predecessor $\widetilde{Q}^j$ of Q^j. By assumption (ii) we have $\widetilde{Q}^j \subset B$ for each j. Hence we obtain

$$A \subset \cup_j \widetilde{Q}^j \subset B.$$

We relabel $\{\widetilde{Q}^j\}$ so that they are nonoverlapping. Therefore we get

$$|A| \leq \sum_i |A \cap \widetilde{Q}^i| \leq \delta \sum_i |\widetilde{Q}^i| \leq \delta|B|.$$

□

5.3. Harnack Inequality

The main result in this section is the following Harnack inequality.

THEOREM 5.10 *Suppose u belongs to $\mathcal{S}(\lambda, \Lambda, f)$ in B_1 with $u \geq 0$ in B_1 for some $f \in C(B_1)$. Then there holds*

$$\sup_{B_{1/2}} u \leq C\Big\{\inf_{B_{1/2}} u + \|f\|_{L^n(B_1)}\Big\}$$

where C is a positive constant depending only on n, λ, and Λ.

The interior Hölder continuity of solutions is a direct consequence, whose proof is identical to that of Corollary 4.18.

COROLLARY 5.11 *Suppose u belongs to $\mathcal{S}(\lambda, \Lambda, f)$ in B_1 for some $f \in C(B_1)$. Then $u \in C^\alpha(B_1)$ for some $\alpha \in (0,1)$ depending only on n, λ, and Λ. Moreover, there holds*

$$|u(x) - u(y)| \le C|x-y|^\alpha \{\sup_{B_1} |u| + \|f\|_{L^n(B_1)}\} \quad \text{for any } x, y \in B_{1/2}$$

where $C = C(n, \lambda, \Lambda)$ is a positive constant.

For convenience we work in cubes instead of balls. We will prove the following result.

LEMMA 5.12 *Suppose u belongs to $\mathcal{S}(\lambda, \Lambda, f)$ in $Q_{4\sqrt{n}}$ with $u \ge 0$ in $Q_{4\sqrt{n}}$ for some $f \in C(Q_{4\sqrt{n}})$. Then there exist two positive constants ε_0 and C, depending only on n, λ, and Λ, such that if $\inf_{Q_{1/4}} u \le 1$ and $\|f\|_{L^n(Q_{4\sqrt{n}})} \le \varepsilon_0$ there holds $\sup_{Q_{1/4}} u \le C$.*

Theorem 5.10 easily follows from Lemma 5.12. For $u \in \mathcal{S}(\lambda, \Lambda, f)$ in $Q_{4\sqrt{n}}$ with $u \ge 0$ in $Q_{4\sqrt{n}}$, consider

$$u_\delta = \frac{u}{\inf_{Q_{1/4}} u + \delta + \frac{1}{\varepsilon_0}\|f\|_{L^n(Q_{4\sqrt{n}})}} \quad \text{for } \delta > 0.$$

We apply Lemma 5.12 to u_δ to get, after letting $\delta \to 0$,

$$\sup_{Q_{1/4}} u \le C\{\inf_{Q_{1/4}} u + \|f\|_{L^n(Q_{4\sqrt{n}})}\}.$$

Then Theorem 5.10 follows by a standard covering argument.

Now we begin to prove Lemma 5.12. The following result is the key ingredient. It claims that if the solution is small somewhere in Q_3 then it is under control in a good portion of Q_1.

LEMMA 5.13 *Suppose u belongs to $\mathcal{S}^+(\lambda, \Lambda, f)$ in $B_{2\sqrt{n}}$ for some $f \in C(B_{2\sqrt{n}})$. Then there exist constants $\varepsilon_0 > 0$, $\mu \in (0,1)$, and $M > 1$, depending only on n, λ, and Λ, such that if*

$$u \ge 0 \text{ in } B_{2\sqrt{n}}, \quad \inf_{Q_3} \inf u \le 1, \quad \|f\|_{L^n(B_{2\sqrt{n}})} \le \varepsilon_0, \tag{5.3}$$

there holds

$$|\{u \le M\} \cap Q_1| > \mu.$$

PROOF: We will construct a function g, which is very concave outside Q_1, such that if we correct u by g the contact set occurs in Q_1. In other words, we localize where contact occurs by choosing suitable functions.

Note $B_{1/4} \subset B_{1/2} \subset Q_1 \subset Q_3 \subset B_{2\sqrt{n}}$. Define g in $B_{2\sqrt{n}}$ by

$$g(x) = -M\left(1 - \frac{|x|^2}{4n}\right)^\beta$$

for large $\beta > 0$ to be determined and some $M > 0$. We choose M, according to β, such that

$$g = 0 \text{ on } \partial B_{2\sqrt{n}} \quad \text{and} \quad g \le -2 \text{ in } Q_3. \tag{5.4}$$

Set $w = u + g$ in $B_{2\sqrt{n}}$. We will show by choosing β large that

$$w \in \mathcal{S}^+(\lambda, \Lambda, f) \quad \text{in } B_{2\sqrt{n}} \setminus Q_1. \tag{5.5}$$

Suppose φ is a quadratic polynomial with the property that $w - \varphi$ has a local minimum at $x_0 \in B_{2\sqrt{n}}$. Then $u - (\varphi - g)$ has a local minimum at $x_0 \in B_{2\sqrt{n}}$. By definitions of $\mathcal{S}^+(\lambda, \Lambda, f)$ and the Pucci extremal operator $\mathcal{M}^-$ we have

$$\mathcal{M}^-(\lambda, \Lambda, D^2\varphi(x_0) - D^2 g(x_0)) \le f(x_0)$$

or

$$\mathcal{M}^-(\lambda, \Lambda, D^2\varphi(x_0)) + \mathcal{M}^-(\lambda, \Lambda, -D^2 g(x_0)) \le f(x_0)$$

where we used the property of $\mathcal{M}^-$. We will choose β large such that

$$\mathcal{M}^-(\lambda, \Lambda, -D^2 g(x_0)) \ge 0 \quad \text{for any } x_0 \in B_{2\sqrt{n}} \setminus B_{1/4}.$$

We need to calculate the Hessian matrix of g. Note

$$D_{ij} g(x) = \frac{M}{2n}\beta\left(1 - \frac{|x|^2}{4n}\right)^{\beta-1}\delta_{ij} - \frac{M}{(2n)^2}\beta(\beta-1)\left(1 - \frac{|x|^2}{4n}\right)^{\beta-2} x_i x_j.$$

If we choose $x = (|x|, 0, \ldots, 0)$ then the eigenvalues of $-D^2 g(x)$ are given by

$$\frac{M}{2n}\beta\left(1 - \frac{|x|^2}{4n}\right)^{\beta-2}\left(\frac{2\beta-1}{4n}|x|^2 - 1\right) \quad \text{with multiplicity } 1,$$

$$-\frac{M}{2n}\beta\left(1 - \frac{|x|^2}{4n}\right)^{\beta-1} \quad \text{with multiplicity } n-1.$$

We choose β large such that for $|x| \ge \frac{1}{4}$ the first eigenvalue is positive and the rest negative, denoted by $e^+(x)$ and $e^-(x)$, respectively. Therefore for $|x| \ge \frac{1}{4}$ we have

$$\begin{aligned}
&\mathcal{M}^-(\lambda, \Lambda, -D^2 g(x)) \\
&\quad = \lambda e^+(x) + (n-1)\Lambda e^-(x) \\
&\quad = \frac{M}{2n}\beta\left(1 - \frac{|x|^2}{4n}\right)^{\beta-2}\left\{\lambda\left(\frac{2\beta-1}{4n}|x|^2 - 1\right) - (n-1)\Lambda\left(1 - \frac{|x|^2}{4n}\right)\right\} \\
&\quad \ge 0
\end{aligned}$$

if we choose β large, depending only on n, λ, and Λ. This finishes the proof of (5.5). In fact, we obtain

$$w \in \mathcal{S}^+(\lambda, \Lambda, f + \eta) \quad \text{in } B_{2\sqrt{n}}$$

for some $\eta \in C_0^\infty(Q_1)$ and $0 \le \eta \le C(n, \lambda, \Lambda)$.

We may apply Theorem 5.8 to w in $B_{2\sqrt{n}}$. Note $\inf_{Q_3} w \le -1$ and $w \ge 0$ on $\partial B_{2\sqrt{n}}$ by (5.3) and (5.4). We obtain

$$1 \le C\Bigg(\int\limits_{B_{2\sqrt{n}} \cap \{w = \Gamma_w\}} (|f| + \eta)^n \Bigg)^{\frac{1}{n}}$$

$$\le C\|f\|_{L^n(B_{2\sqrt{n}})} + C|\{w = \Gamma_w\} \cap Q_1|^{\frac{1}{n}}.$$

Choosing ε_0 small enough we get

$$\frac{1}{2} \le C|\{w = \Gamma_w\} \cap Q_1|^{\frac{1}{n}} \le C|\{u \le M\} \cap Q_1|^{\frac{1}{n}}$$

since $w(x) = \Gamma_w(x)$ implies $w(x) \le 0$ and hence $u(x) \le -g(x) \le M$. This finishes the proof. □

Next we prove the power decay of distribution functions.

LEMMA 5.14 *Let u belong to $S^+(\lambda, \Lambda, f)$ in $B_{2\sqrt{n}}$ for some $f \in C(B_{2\sqrt{n}})$. Then there exist positive constants ε_0, ε, and C, depending only on n, λ, and Λ, such that if*

$$u \ge 0 \text{ in } B_{2\sqrt{n}}, \quad \inf_{Q_3} \inf u \le 1, \quad \|f\|_{L^n(B_{2\sqrt{n}})} \le \varepsilon_0, \tag{5.6}$$

there holds

$$|\{u \ge t\} \cap Q_1| \le Ct^{-\varepsilon} \quad \text{for } t > 0.$$

PROOF: We will prove that under assumption (5.6) there holds

$$|\{u > M^k\} \cap Q_1| \le (1 - \mu)^k \quad \text{for } k = 1, 2, \ldots, \tag{5.7}$$

where M and μ are as in Lemma 5.13.

For $k = 1$, (5.7) is just Lemma 5.13. Suppose now (5.7) holds for $k - 1$. Set

$$A = \{u > M^k\} \cap Q_1, \quad B = \{u > M^{k-1}\} \cap Q_1.$$

We will use Lemma 5.9 to prove that

$$|A| \le (1 - \mu)|B|. \tag{5.8}$$

Clearly $A \subset B \subset Q_1$ and $|A| \le |\{u > M\} \cap Q_1| \le 1 - \mu$ by Lemma 5.13. We claim that if $Q = Q_r(x_0)$ is a cube in Q_1 such that

$$|A \cap Q| > (1 - \mu)|Q| \tag{5.9}$$

then $\tilde{Q} \cap Q_1 \subset B$ for $\tilde{Q} = Q_{3r}(x_0)$.

We prove it by contradiction. Suppose not. We may take $\tilde{x} \in \tilde{Q}$ such that $u(\tilde{x}) \le M^{k-1}$. Consider the transformation

$$x = x_0 + ry \quad \text{for } y \in Q_1 \text{ and } x \in Q = Q_r(x_0)$$

and the function

$$\tilde{u}(y) = \frac{1}{M^{k-1}} u(x).$$

Then $\widetilde{u} \geq 0$ in $B_{2\sqrt{n}}$ and $\inf_{Q_3} \widetilde{u} \leq 1$. It is easy to check that $\widetilde{u} \in \mathcal{S}^+(\lambda, \Lambda, \widetilde{f})$ in $B_{2\sqrt{n}}$ with $\|\widetilde{f}\|_{L^n(B_{2\sqrt{n}})} \leq \varepsilon_0$. In fact, we have

$$\widetilde{f}(y) = \frac{r^2}{M^{k-1}} f(x) \quad \text{for } y \in B_{2\sqrt{n}}$$

and hence

$$\|\widetilde{f}\|_{L^n(B_{2\sqrt{n}})} \leq \frac{r}{M^{k-1}} \|f\|_{L^n(B_{2\sqrt{n}})} \leq \|f\|_{L^n(B_{2\sqrt{n}})} \leq \varepsilon_0.$$

Hence $\widetilde{u}$ satisfies assumption (5.6). We may apply Lemma 5.13 to $\widetilde{u}$ to get

$$\mu < |\{\widetilde{u}(y) \leq M\} \cap Q_1| = r^{-n} |\{u(x) \leq M^k\} \cap Q|.$$

Hence $|Q \cap A^{c}| > \mu |Q|$, which contradicts (5.9). We are in a position to apply Lemma 5.9 to get (5.8). $\square$

PROOF OF LEMMA 5.12: We prove that there exist two constants $\theta > 1$ and $M_0 \gg 1$, depending only on n, λ, and Λ, such that if $u(x_0) = P > M_0$ for some $x_0 \in B_{1/4}$ there exists a sequence $\{x_k\} \in B_{1/2}$ such that

$$u(x_k) \geq \theta^k P \quad \text{for } k = 0, 1, \ldots.$$

This contradicts the boundedness of u, hence we conclude that $\sup_{B_{1/4}} u \leq M_0$.

Suppose $u(x_0) = P > M_0$ for some $x_0 \in B_{1/4}$. We will determine M_0 and θ in the process. Consider a cube $Q_r(x_0)$, centered at x_0 with side length r, which will be chosen later. We want to find a point $x_1 \in Q_{4\sqrt{n}r}(x_0)$ such that $u(x_1) \geq \theta P$. To do that we first choose r such that $\{u > \frac{P}{2}\}$ covers less than half of $Q_r(x_0)$. This can be done by using the power decay of the distribution function of u.

Note $\inf_{Q_3} u \leq \inf_{Q_{1/4}} u \leq 1$. Hence Lemma 5.14 implies

$$\left|\left\{u > \frac{P}{2}\right\} \cap Q_1\right| \leq C\left(\frac{P}{2}\right)^{-\varepsilon}.$$

We choose r such that $\frac{r^n}{2} \geq C(\frac{P}{2})^{-\varepsilon}$ and $r \leq \frac{1}{4}$. Hence we have, for such r, $Q_r(x_0) \subset Q_1$ and

$$\frac{1}{|Q_r(x_0)|}\left|\left\{u > \frac{P}{2}\right\} \cap Q_r(x_0)\right| \leq \frac{1}{2}. \tag{5.10}$$

Next we show that for $\theta > 1$, with $\theta - 1$ small, $u \geq \theta P$ at some point in $Q_{4\sqrt{n}r}(x_0)$. We prove it by contradiction. Suppose $u \leq \theta P$ in $Q_{4\sqrt{n}r}(x_0)$. Consider the transformation

$$x = x_0 + ry \quad \text{for } y \in Q_{4\sqrt{n}} \text{ and } x \in Q_{4\sqrt{n}r}(x_0)$$

and the function

$$\widetilde{u}(y) = \frac{\theta P - u(x)}{(\theta - 1)P}.$$

Obviously $\tilde{u} \geq 0$ in $B_{2\sqrt{n}}$ and $\tilde{u}(0) = 1$, hence $\inf_{Q_3} \tilde{u} \leq 1$. It is easy to check that $\tilde{u} \in \mathcal{S}^+(\lambda, \Lambda, \tilde{f})$ in $B_{2\sqrt{n}}$ with $\|\tilde{f}\|_{L^n(B_{2\sqrt{n}})} \leq \varepsilon_0$. In fact, we have

$$\tilde{f}(y) = -\frac{r^2}{(\theta - 1)P} f(x) \quad \text{for } y \in B_{2\sqrt{n}}$$

and hence

$$\|\tilde{f}\|_{L^n(B_{2\sqrt{n}})} \leq \frac{r}{(\theta - 1)P} \|f\|_{L^n(B_{2\sqrt{n}r}(x_0))} \leq \varepsilon_0$$

if we choose P such that $r \leq (\theta - 1)P$. Hence we may apply Lemma 5.13 to $\tilde{u}$. Note that $u(x) \leq \frac{P}{2}$ if and only if $\tilde{u}(y) \geq \frac{\theta - 1/2}{\theta - 1}$ and that $\frac{\theta - 1/2}{\theta - 1}$ is large if θ is close to 1. So we obtain

$$\begin{aligned} \frac{1}{|Q_r(x_0)|} \left| \left\{ u \leq \frac{P}{2} \right\} \cap Q_r(x_0) \right| &= \left| \left\{ \tilde{u} \geq \frac{\theta - \frac{1}{2}}{\theta - 1} \right\} \cap Q_1 \right| \\ &\leq C \left(\frac{\theta - \frac{1}{2}}{\theta - 1} \right)^{-\varepsilon} < \frac{1}{2} \end{aligned}$$

if θ is chosen close to 1. This contradicts (5.10).

Hence we conclude that there exists a $\theta = \theta(n, \lambda, \Lambda) > 1$ such that if

$$u(x_0) = P \quad \text{for some } x_0 \in B_{1/4}$$

then

$$u(x_1) \geq \theta P \quad \text{for some } x_1 \in Q_{4\sqrt{n}r}(x_0) \subset B_{2nr}(x_0)$$

provided

$$C(n, \lambda, \Lambda) P^{-\frac{\varepsilon}{n}} \leq r \leq (\theta - 1)P.$$

So we need to choose P such that $P \geq (\frac{C}{\theta - 1})^{n/(n+\varepsilon)}$ and then take $r = CP^{-\varepsilon/n}$.

Now we may iterate the above result to get a sequence $\{x_k\}$ such that for any $k = 1, 2, \ldots,$

$$u(x_k) \geq \theta^k P \quad \text{for some } x_k \in B_{2nr_k}(x_{k-1})$$

where $r_k = C(\theta^{k-1}P)^{-\varepsilon/n} = C\theta^{-(k-1)\varepsilon/n}P^{-\varepsilon/n}$. In order to have $\{x_k\} \in B_{1/2}$ we need $\sum 2nr_k < \frac{1}{4}$. Hence we choose M_0 such that

$$M_0^{\varepsilon/n} \geq 8nC \sum_{k=1}^{\infty} \theta^{-(k-1)\frac{\varepsilon}{n}} \quad \text{and} \quad M_0 \geq \left(\frac{C}{\theta - 1} \right)^{\frac{n}{n+\varepsilon}}$$

and then take $P > M_0$. This finishes the proof. □

In the rest of this section we prove a technical lemma concerning the second-order derivatives of functions in $\mathcal{S}(\lambda, \Lambda, f)$. Such results will be needed in the discussion of $W^{2,p}$-estimates. First we introduce some terminology.

Let Ω be a bounded domain and u be a continuous function in Ω. We define for $M > 0$

$$G_M^-(u,\Omega) = \{x_0 \in \Omega : \text{there exists an affine function } L \text{ such that } L(x) - \tfrac{M}{2}|x-x_0|^2 \le u(x) \text{ for } x \in \Omega \text{ with equality at } x_0\},$$

$$G_M^+(u,\Omega) = \{x_0 \in \Omega : \text{there exists an affine function } L \text{ such that } L(x) + \tfrac{M}{2}|x-x_0|^2 \ge u(x) \text{ for } x \in \Omega \text{ with equality at } x_0\},$$

$$G_M(u,\Omega) = G_M^+(u,\Omega) \cap G_M^-(u,\Omega).$$

We also define

$$\begin{aligned} A_M^-(u,\Omega) &= \Omega \setminus G_M^-(u,\Omega), \\ A_M^+(u,\Omega) &= \Omega \setminus G_M^+(u,\Omega), \\ A_M(u,\Omega) &= \Omega \setminus G_M(u,\Omega). \end{aligned}$$

In other words, $G_M^-(u,\Omega)$ (respectively, $G_M^+(u,\Omega)$) consists of points where there is a concave (respectively, convex) paraboloid of opening M touching u from below (respectively, above). Intuitively $|A_M(u,\Omega)|$ behaves like the distribution function of D^2u. Hence for integrability of D^2u we need to study the decay of $|A_M(u,\Omega)|$.

LEMMA 5.15 *Suppose that Ω is a bounded domain with $B_{6\sqrt{n}} \subset \Omega$ and that u belongs to $S^+(\lambda,\Lambda,f)$ in $B_{6\sqrt{n}}$ for some $f \in C(B_{6\sqrt{n}})$. Then there exist positive constants δ_0, μ, and C, depending only on n, λ, and Λ, such that if $|u| \le 1$ in Ω and $\|f\|_{L^n(B_{6\sqrt{n}})} \le \delta_0$ there holds*

$$|A_t^-(u,\Omega) \cap Q_1| \le Ct^{-\mu} \quad \text{for any } t > 0.$$

If, in addition, $u \in S(\lambda,\Lambda,f)$ in $B_{6\sqrt{n}}$, then

$$|A_t(u,\Omega) \cap Q_1| \le Ct^{-\mu} \quad \text{for any } t > 0.$$

In the proof of Lemma 5.15 we need the *maximal functions* of local integrable functions. For $g \in L^1_{\text{loc}}(\mathbb{R}^n)$ we define

$$m(g)(x) = \sup_{r>0} \frac{1}{|Q_r(x)|} \int_{Q_r(x)} |g|.$$

The maximal operator m is of weak type $(1,1)$ and of strong type (p,p) for $1 < p \le \infty$, that is,

$$\begin{aligned} |\{x \in \mathbb{R}^n : m(g)(x) \ge t\}| &\le \frac{c_1(n)}{t}\|g\|_{L^1(\mathbb{R}^n)} && \text{for any } t > 0, \\ \|m(g)\|_{L^p(\mathbb{R}^n)} &\le c_2(n,p)\|g\|_{L^p(\mathbb{R}^n)} && \text{for } 1 < p \le \infty. \end{aligned}$$

Now we begin to prove Lemma 5.15. The following result is the key ingredient. It claims that if u has a tangent paraboloid with opening 1 from below somewhere in Q_3, then the set where u has a tangent paraboloid from below with opening M in Q_1 is large. Compare it with Lemma 5.13.

LEMMA 5.16 *Suppose that Ω is a bounded domain with $B_{6\sqrt{n}} \subset \Omega$ and that u belongs to $S^+(\lambda, \Lambda, f)$ in $B_{6\sqrt{n}}$ for some $f \in C(B_{6\sqrt{n}})$. Then there exist constants $0 < \sigma < 1$, $\delta_0 > 0$, and $M > 1$, depending only on n, λ, and Λ, such that if $\|f\|_{L^n(B_{6\sqrt{n}})} \leq \delta_0$ and $G_1^-(u, \Omega) \cap Q_3 \neq \emptyset$, then*

$$|G_M^-(u, \Omega) \cap Q_1| \geq 1 - \sigma.$$

PROOF: Since $G_1^-(u, \Omega) \cap Q_3 \neq \emptyset$, there is an affine function L_1 such that

$$v \geq P_1 \quad \text{in } \Omega \text{ with equality at some point in } Q_3$$

where

$$v(x) = \frac{u(x)}{2n} + L_1(x) \quad \text{and} \quad P_1(x) = 1 - \frac{|x|^2}{4n}.$$

This implies $v \geq 0$ in $B_{2\sqrt{n}}$ and $\inf_{Q_3} \leq 1$. Then as in the proof of Lemma 5.13, for $w = v + g$, where g is the function constructed in Lemma 5.13, we have

$$|\{w = \Gamma_w\} \cap Q_1| \geq 1 - \sigma$$

for some $\sigma \in (0, 1)$ if δ_0 is chosen small. Now we need to prove $\{w = \Gamma_w\} \cap Q_1 \subset G_M^-(u, \Omega) \cap Q_1$ for some $M > 1$. Let $x_0 \in \{w = \Gamma_w\} \cap Q_1$ and take an affine function L_2 with $L_2 < 0$ on $\partial B_{2\sqrt{n}}$ and

$$L_2 \leq \Gamma_w \leq v + g \quad \text{in } B_{2\sqrt{n}} \text{ with equality at } x_0.$$

It follows that

$$P_2 \leq L_2 - g \leq v \quad \text{in } B_{2\sqrt{n}} \text{ with equality at } x_0 \tag{5.11}$$

for a concave paraboloid P_2 of opening $M_0 = M_0(n, \lambda, \Lambda) > 0$.

Next we prove $P_2 \leq v$ in $\Omega \setminus B_{2\sqrt{n}}$. Note that $P_2 < -g = 0 = P_1$ on $\partial B_{2\sqrt{n}}$ and that $P_2(x_0) = v(x_0) \geq P_1(x_0)$ with $x_0 \in Q_1 \subset B_{2\sqrt{n}}$. If we take $M_0 > \frac{1}{2n}$, then $\{P_2 - P_1 \geq 0\}$ is convex. We conclude that $P_2 - P_1 < 0$ in $\mathbb{R}^n \setminus B_{2\sqrt{n}}$. Hence we have $P_2 \leq P_1 \leq v$ in $\Omega \setminus B_{2\sqrt{n}}$. By (5.11) and the definition of v, we get $x_0 \in G_{2nM_0}^-(u, \Omega) \cap Q_1$ with $2nM_0 > 1$. □

PROOF OF LEMMA 5.15: Recall $B_{6\sqrt{n}} \subset \Omega$,$u \in S^+(\lambda, \Lambda, f)$ in $B_{6\sqrt{n}}$ and

$$|u|_{L^\infty(\Omega)} \leq 1, \quad \|f\|_{L^n(B_{6\sqrt{n}})} \leq \delta_0. \tag{5.12}$$

We will prove that there exist constants $M > 1$ and $0 < \gamma < 1$, depending only on n, λ, and Λ, such that

$$|A_{M^k}^-(u, \Omega) \cap Q_1| \leq \gamma^k \quad \text{for any } k = 0, 1, \ldots.$$

Step 1. There exist constants $M > 1$ and $0 < \sigma < 1$ such that

$$|G_M^-(u,\Omega) \cap Q_1| \geq 1 - \sigma. \tag{5.13}$$

It is easy to see that $|u|_{L^\infty(\Omega)} \leq 1$ implies that

$$G_{c(n)}^-(u,\Omega) \cap Q_3 \neq \varnothing$$

for some constant $c(n)$ depending only on n. We may apply Lemma 5.15 to $u/c(n)$ to get (5.13). By a simple adjustment we may assume that δ_0, M, and σ in Step 1 are the same as those in Lemma 5.15.

Step 2. We extend f by 0 outside $B_{6\sqrt{n}}$ and set for $k = 0, 1, \dots,$

$$A = A_{M^{k+1}}^-(u,\Omega) \cap Q_1,$$
$$B = (A_{M^k}^-(u,\Omega) \cap Q_1) \cup \{x \in Q_1 : m(f^n)(x) \geq (c_1 M^k)^n\},$$

for some $c_1 > 0$ to be determined. Then there holds

$$|A| \leq \sigma |B|$$

where $M > 1$ and $0 < \sigma < 1$ are as before. Recall that $m(f^n)$ denotes the maximal function of f^n.

We prove this by Lemma 5.9. It is easy to see that $|A| \leq \sigma$ since we have $|G_{M^{k+1}}^-(u,\Omega) \cap Q_1| \geq |G_M^-(u,\Omega) \cap Q_1| \geq 1-\sigma$ by Step 1.

Next we claim that if $Q = Q_r(x_0)$ is a cube in Q_1 such that

$$|A_{M^{k+1}}^-(u,\Omega) \cap Q| = |A \cap Q| > \sigma |Q|, \tag{5.14}$$

then $\widetilde{Q} \cap Q_1 \subset B$ for $\widetilde{Q} = Q_{3r}(x_0)$.

We prove this by contradiction. Suppose not. We may take an $\widetilde{x}$ such that

$$\widetilde{x} \in G_{M^k}^-(u,\Omega) \cap \widetilde{Q} \quad \text{and} \quad \sup_{r>0} \frac{1}{|Q_r(\widetilde{x})|} \int_{Q_r(\widetilde{x})} |f|^n \leq (c_1 M^k)^n.$$

Consider the transformation

$$x = x_0 + ry \quad \text{for } y \in Q_1 \text{ and } x \in Q = Q_r(x_0)$$

and the function

$$\widetilde{u}(y) = \frac{1}{r^2 M^k} u(x).$$

It is easy to check that $B_{6\sqrt{n}} \subset \widetilde{\Omega}$, the image of Ω under the transformation above, and that $\widetilde{u} \in \mathcal{S}^+(\lambda, \Lambda, \widetilde{f})$ in $B_{6\sqrt{n}}$ with

$$\widetilde{f}(y) = \frac{1}{M^k} f(x) \quad \text{for } y \in B_{6\sqrt{n}}.$$

By the choice of $\widetilde{x}$ we have

$$G_1^-(\widetilde{u}, \widetilde{\Omega}) \cap Q_3 \neq \varnothing.$$

Since $B_{6\sqrt{n}r}(x_0) \subset Q_{15\sqrt{n}r}(\widetilde{x})$ there holds

$$\|\widetilde{f}\|_{L^n(B_{6\sqrt{n}})} \leq \frac{1}{rM^k} \|f\|_{L^n(Q_{15\sqrt{n}r}(\widetilde{x}))} \leq c(n)c_1 \leq \delta_0$$

if we take c_1 small enough, depending only on n, λ, and Λ.

Hence $\widetilde{u}$ satisfies the assumption of Lemma 5.16 with Ω replaced by $\widetilde{\Omega}$. We may apply Lemma 5.16 to $\widetilde{u}$ to get

$$|G_M^-(\widetilde{u}, \widetilde{\Omega}) \cap Q_1| \geq 1 - \sigma \quad \text{or} \quad |G_{M^{k+1}}^-(u, \Omega) \cap Q| > (1 - \sigma)|Q|.$$

This contradicts (5.14). We are in a position to apply Lemma 5.9.

Step 3. We finish the proof of Lemma 5.15. Define for $k = 0, 1, \ldots,$

$$\alpha_k = |A_{M^k}^-(u, \Omega) \cap Q_1|,$$

$$\beta_k = |\{x \in Q_1 : m(f^n)(x) \geq (c_1 M^k)^n\}|.$$

Then Step 2 implies $\alpha_{k+1} \leq \sigma(\alpha_k + \beta_k)$ for any $k = 0, 1, \ldots$. Hence by iteration we have

$$\alpha_k \leq \sigma^k + \sum_{i=0}^{k-1} \sigma^{k-i} \beta_i.$$

Since $\|f^n\|_{L^1} \leq \delta_0^n$ and the maximal operator is of weak type (1, 1), we conclude that

$$\beta_k \leq c(n)\delta_0^n (c_1 M^k)^{-n} = C(n, \lambda, \Lambda) M^{-nk}.$$

This implies

$$\sum_{i=0}^{k-1} \sigma^{k-i} \beta_i \leq C \sum_{i=0}^{k-1} \sigma^{k-i} M^{-ni} \leq C k \gamma_0^k$$

with $\gamma_0 = \max\{\sigma, M^{-n}\} < 1$. Therefore we obtain for k large

$$\alpha_k \leq \sigma^k + C k \gamma_0^k \leq (1 + Ck)\gamma_0^k \leq \gamma^k$$

for some $\gamma = \gamma(n, \lambda, \Lambda) \in (0, 1)$.

This finishes the proof. □

REMARK 5.17. The polynomial decay of the function

$$\mu(t) = |A_t(u, \Omega) \cap Q_1|$$

for $u \in \mathcal{S}(\lambda, \Lambda, f)$ implies that $D^2 u$ is L^p-integrable in Q_1 for small $p > 0$ depending only on n, λ, and Λ. In order to show the L^p-integrability for large p we need to speed up the convergence in the proof of Lemma 5.15. We will discuss $W^{2,p}$-estimates in Section 5.5.

5.4. Schauder Estimates

In this section we will prove the Schauder estimates for viscosity solutions. Throughout this section we always assume that $a_{ij} \in C(B_1)$ satisfies

$$\lambda|\xi|^2 \leq a_{ij}(x)\xi_i\xi_j \leq \Lambda|\xi|^2 \quad \text{for any } x \in B_1 \text{ and any } \xi \in \mathbb{R}^n$$

for some positive constants λ and Λ and that f is a continuous function in B_1.

The following approximation result plays an important role in the discussion of regularity theory.

LEMMA 5.18 *Suppose $u \in C(B_1)$ is a viscosity solution of*

$$a_{ij} D_{ij} u = f \quad \text{in } B_1$$

with $|u| \le 1$ in B_1. Assume for some $0 < \varepsilon < \frac{1}{16}$,

$$\|a_{ij} - a_{ij}(0)\|_{L^n(B_{3/4})} \le \varepsilon.$$

Then there exists a function $h \in C(\bar{B}_{3/4})$ with $a_{ij}(0) D_{ij} h = 0$ in $B_{3/4}$ and $|h| \le 1$ in $B_{3/4}$ such that

$$|u - h|_{L^\infty(B_{1/2})} \le C\{\varepsilon^\gamma + \|f\|_{L^n(B_1)}\}$$

where $C = C(n, \lambda, \Lambda)$ is a positive constant and $\gamma = \gamma(n, \lambda, \Lambda) \in (0, 1)$.

PROOF: Solve for $h \in C(\bar{B}_{3/4}) \cap C^\infty(B_{3/4})$ such that

$$a_{ij}(0) D_{ij} h = 0 \quad \text{in } B_{3/4},$$
$$h = u \quad \text{on } \partial B_{3/4}.$$

The maximum principle implies $|h| \le 1$ in $B_{3/4}$. Note that u belongs to $\mathcal{S}(\lambda, \Lambda, f)$ in B_1. Corollary 5.11 implies that $u \in C^\alpha(\bar{B}_{3/4})$ for some $\alpha = \alpha(n, \lambda, \Lambda) \in (0, 1)$ with the estimate

$$\|u\|_{C^\alpha(\bar{B}_{3/4})} \le C(n, \lambda, \Lambda)\{1 + \|f\|_{L^n(B_1)}\}.$$

By Lemma 1.35 we have

$$\|h\|_{C^{\alpha/2}(\bar{B}_{3/4})} \le C\|u\|_{C^\alpha(\bar{B}_{3/4})} \le C(n, \lambda, \Lambda)\{1 + \|f\|_{L^n(B_1)}\}.$$

Since $u - h = 0$ on $\partial B_{3/4}$ we get for any $0 < \delta < \frac{1}{4}$

$$|u - h|_{L^\infty(\partial B_{3/4-\delta})} \le C\delta^{\alpha/2}\{1 + \|f\|_{L^n(B_1)}\}. \tag{5.15}$$

We claim for any $0 < \delta < 1$

$$|D^2 h|_{L^\infty(B_{3/4-\delta})} \le C\delta^{\alpha/2-2}\{1 + \|f\|_{L^n(B_1)}\}. \tag{5.16}$$

In fact, for any $x_0 \in B_{3/4-\delta}$ we apply interior C^2-estimate to $h - h(x_1)$ in $B_\delta(x_0) \subset B_{3/4}$ for some $x_1 \in \partial B_\delta(x_0)$ and obtain

$$|D^2 h(x_0)| \le C\delta^{-2} \sup_{B_\delta(x_0)} |h - h(x_1)| \le C\delta^{-2}\delta^{\alpha/2}\{1 + \|f\|_{L^n(B_1)}\}.$$

Note that $u - h$ is a viscosity solution of

$$a_{ij} D_{ij}(u - h) = f - (a_{ij} - a_{ij}(0)) D_{ij} h \equiv F \quad \text{in } B_{3/4}.$$

By Theorem 5.8 (the Alexandroff maximum principle) we have with (5.15) and (5.16)

$$\begin{aligned}
&|u - h|_{L^\infty(B_{3/4-\delta})} \\
&\quad \le |u - h|_{L^\infty(\partial B_{3/4-\delta})} + C\|F\|_{L^n(B_{3/4-\delta})} \\
&\quad \le |u - h|_{L^\infty(\partial B_{3/4-\delta})} \\
&\qquad + C|D^2 h|_{L^\infty(B_{3/4-\delta})}\|a_{ij} - a_{ij}(0)\|_{L^n(B_{3/4})} + C\|f\|_{L^n(B_1)} \\
&\quad \le C(\delta^{\alpha/2} + \delta^{\alpha/2-2}\varepsilon)\{1 + \|f\|_{L^n(B_1)}\} + C\|f\|_{L^n(B_1)}.
\end{aligned}$$

Take $\delta = \varepsilon^{1/2} < \frac{1}{4}$ and then $\gamma = \frac{\alpha}{4}$. This finishes the proof. □

For the next result we need to introduce the following concept.

DEFINITION 5.19 A function g is Hölder-continuous at 0 with exponent α in the L^n sense if

$$[g]_{C^\alpha_{L^n}}(0) \equiv \sup_{0<r<1} \frac{1}{r^\alpha}\left(\frac{1}{|B_r|}\int_{B_r} |g - g(0)|^n\right)^{\frac{1}{n}} < \infty.$$

Now we state the Schauder estimates.

THEOREM 5.20 *Suppose $u \in C(B_1)$ is a viscosity solution of*

$$a_{ij} D_{ij} u = f \quad \textit{in } B_1.$$

Assume $\{a_{ij}\}$ is Hölder-continuous at 0 with exponent α in the L^n sense for some $\alpha \in (0,1)$. If f is Hölder-continuous at 0 with exponent α in the L^n-sense, then u is $C^{2,\alpha}$ at 0. Moreover, there exists a polynomial P of degree 2 such that

$$|u - P|_{L^\infty(B_r(0))} \le C_* r^{2+\alpha} \quad \textit{for any } 0 < r < 1,$$
$$|P(0)| + |DP(0)| + |D^2P(0)| \le C_*,$$
$$C_* \le C\{|u|_{L^\infty(B_1)} + |f(0)| + [f]_{C^\alpha_{L^n}}(0)\}$$

where C is a positive constant depending only on n, λ, Λ, α, and $[a_{ij}]_{C^\alpha_{L^n}}(0)$.

PROOF: First we assume $f(0) = 0$. For that we may consider $v = u - b_{ij}x_i x_j f(0)/2$ for some constant matrix $\{b_{ij}\}$ such that $a_{ij}(0)b_{ij} = 1$. By scaling we also assume that $[a_{ij}]_{C^\alpha_{L^n}}(0)$ is small. Next by considering for $\delta > 0$

$$\frac{u}{|u|_{L^\infty(B_1)} + \frac{1}{\delta}[f]_{C^\alpha_{L^n}}(0)}$$

we may assume $|u|_{L^\infty(B_1)} \le 1$ and $[f]_{C^\alpha_{L^n}}(0) \le \delta$.

In the following we prove that there is a constant $\delta > 0$, depending only on n, λ, Λ, and α, such that if $u \in C(B_1)$ is a viscosity solution of

$$a_{ij} D_{ij} u = f \quad \text{in } B_1$$

with

$$|u|_{L^\infty(B_1)} \le 1, \quad [a_{ij}]_{C^\alpha_{L^n}}(0) \le \delta, \quad \left(\frac{1}{|B_r|}\int_{B_r} |f|^n\right)^{\frac{1}{n}} \le \delta r^\alpha \text{ for any } 0 < r < 1,$$

then there exists a polynomial P of degree 2 such that

$$|u - P|_{L^\infty(B_r(0))} \le C r^{2+\alpha} \quad \text{for any } 0 < r < 1 \tag{5.17}$$

and

$$|P(0)| + |DP(0)| + |D^2P(0)| \le C \tag{5.18}$$

for some positive constant C depending only on n, λ, Λ, and α.

We claim that there exist $0 < \mu < 1$, depending only on n, λ, Λ, and α, and a sequence of polynomials of degree 2

$$P_k(x) = a_k + b_k \cdot x + \frac{1}{2} x^{\mathsf{T}} C_k x$$

such that for any $k = 0, 1, \ldots,$

$$a_{ij}(0) D_{ij} P_k = 0, \quad |u - P_k|_{L^\infty(B_{\mu^k})} \leq \mu^{k(2+\alpha)}, \tag{5.19}$$

$$\begin{aligned} &|a_k - a_{k-1}| + \mu^{k-1}|b_k - b_{k-1}| + \mu^{2(k-1)}|C_k - C_{k-1}| \\ &\leq C\mu^{(k-1)(2+\alpha)} \end{aligned} \tag{5.20}$$

where $P_0 = P_{-1} \equiv 0$ and C is a positive constant depending only on n, λ, Λ, and α.

We first prove that Theorem 5.20 follows from (5.19) and (5.20). It is easy to see that a_k, b_k, and C_k converge and that the limiting polynomial

$$p(x) = a_\infty + b_\infty \cdot x + \frac{1}{2} x^{\mathsf{T}} C_\infty x$$

satisfies

$$|P_k(x) - p(x)| \leq C\{|x|^2 \mu^{\alpha k} + |x|\mu^{(\alpha+1)k} + \mu^{(\alpha+2)k}\} \leq C\mu^{(2+\alpha)k}$$

for any $|x| \leq \mu^k$. Hence we have for $|x| \leq \mu^k$

$$|u(x) - p(x)| \leq |u(x) - P_k(x)| + |P_k(x) - p(x)| \leq C\mu^{(2+\alpha)k},$$

which implies that

$$|u(x) - p(x)| \leq C|x|^{2+\alpha} \quad \text{for any } x \in B_1.$$

Now we prove (5.19) and (5.20). Clearly (5.19) and (5.20) hold for $k = 0$. Assume they hold for $k = 0, 1, \ldots, l$. We prove for $k = l + 1$. Consider the function

$$\widetilde{u}(y) = \frac{1}{\mu^{l(2+\alpha)}}(u - P_l)(\mu^l y) \quad \text{for } y \in B_1.$$

Then $\widetilde{u} \in C(B_1)$ is a viscosity solution of

$$\widetilde{a}_{ij} D_{ij} \widetilde{u} = \widetilde{f} \quad \text{in } B_1$$

with

$$\begin{aligned} \widetilde{a}_{ij}(y) &= \frac{1}{\mu^{l\alpha}} a_{ij}(\mu^l y), \\ \widetilde{f}(y) &= \frac{1}{\mu^{l\alpha}}\{f(\mu^l y) - a_{ij}(\mu^l y) D_{ij} P_k\}. \end{aligned}$$

Now we check that $\widetilde{u}$ satisfies the assumptions of Lemma 5.18. For that we calculate

$$\|\widetilde{a}_{ij} - \widetilde{a}_{ij}(0)\|_{L^n(B_1)} \leq \frac{1}{\mu^{l\alpha}}\|a_{ij} - a_{ij}(0)\|_{L^n(B_{\mu^l})} \leq [a_{ij}]_{C^\alpha_{L^n}}(0) \leq \delta$$

and

$$\|\tilde{f}\|_{L^n(B_1)} \le \frac{1}{\mu^{l\alpha}}\|f\|_{L^n(B_{\mu^l})} + \frac{1}{\mu^{l\alpha}}\sup|D^2P_l|\|a_{ij}-a_{ij}(0)\|_{L^n(B_{\mu^l})} \le \delta + C\delta$$

where we used

$$|D^2P_l| \le \sum_{k=1}^{l}|D^2P_k - D^2P_{k-1}| \le \sum_{k=1}^{l}\mu^{(k-1)\alpha} \le C.$$

Hence we take $\varepsilon = C(n,\lambda,\Lambda)\delta$ in Lemma 5.18. Then by Lemma 5.18 there exists a function $h \in C(\bar{B}_{3/4})$ with $\tilde{a}_{ij}(0)D_{ij}h = 0$ in $B_{3/4}$ and $|h| \le 1$ in $B_{3/4}$ such that

$$|\tilde{u} - h|_{L^\infty(B_{1/2})} \le C\{\varepsilon^\gamma + \varepsilon\} \le 2C\varepsilon^\gamma.$$

Write $\tilde{P}(y) = h(0) + Dh(0) + y^{\mathrm{T}}D^2h(0)y/2$. Then by interior estimates for h we have

$$|\tilde{u} - \tilde{P}|_{L^\infty(B_\mu)} \le |\tilde{u} - h|_{L^\infty(B_\mu)} + |h - \tilde{P}|_{L^\infty(B_\mu)} \le 2C\varepsilon^\gamma + C\mu^3 \le \mu^{2+\alpha}$$

by choosing μ small and then ε small accordingly. Rescaling back, we have

$$|u(x) - P_l(x) - \mu^{l(2+\alpha)}\tilde{P}(\mu^{-l}x)| \le \mu^{(l+1)(2+\alpha)} \quad \text{for any } x \in B_{\mu^{l+1}}.$$

This implies (5.19) for $k = l + 1$ if we define

$$P_{k+1}(x) = P_k(x) + \mu^{l(2+\alpha)}\tilde{P}(\mu^{-l}x).$$

Estimate (5.20) follows easily. □

To finish this section we state the Cordes-Nirenberg type estimate. The proof is similar to that of Theorem 5.20.

THEOREM 5.21 *Suppose $u \in C(B_1)$ is a viscosity solution of*

$$a_{ij}D_{ij}u = f \quad \textit{in } B_1.$$

Then for any $\alpha \in (0,1)$ there exists an $\theta > 0$, depending only on n, λ, Λ, and α, such that if

$$\left(\frac{1}{|B_r|}\int_{B_r}|a_{ij} - a_{ij}(0)|^n\right)^{\frac{1}{n}} \le \theta \quad \textit{for any } 0 < r \le 1,$$

then u is $C^{1,\alpha}$ at 0; that is, there exists an affine function L such that

$$|u - L|_{L^\infty(B_r(0))} \le C_* r^{1+\alpha} \quad \textit{for any } 0 < r < 1,$$

$$|L(0)| + |DL(0)| \le C_*,$$

$$C_* \le C\left\{|u|_{L^\infty(B_1)} + \sup_{0<r<1} r^{1-\alpha}\left(\frac{1}{|B_r|}\int_{B_r}|f|^n\right)^{\frac{1}{n}}\right\},$$

where C is a positive constant depending only on n, λ, Λ, and α.

5.5. $W^{2,p}$ Estimates

In this section we will prove the $W^{2,p}$-estimates for viscosity solutions. We always assume throughout this section that $a_{ij} \in C(B_1)$ satisfies

$$\lambda|\xi|^2 \leq a_{ij}(x)\xi_i\xi_j \leq \Lambda|\xi|^2 \quad \text{for any } x \in B_1 \text{ and any } \xi \in \mathbb{R}^n$$

for some positive constants λ and Λ and that f is a continuous function in B_1.

The main result in this section is the following theorem.

THEOREM 5.22 *Suppose $u \in C(B_1)$ is a viscosity solution of*

$$a_{ij}D_{ij}u = f \quad \text{in } B_1.$$

Then for any $p \in (n, \infty)$ there exists an $\varepsilon > 0$, depending only on n, λ, Λ, and p, such that if

$$\left(\frac{1}{|B_r(x_0)|}\int_{B_r(x_0)} |a_{ij} - a_{ij}(x_0)|^n\right)^{\frac{1}{n}} \leq \varepsilon \quad \text{for any } B_r(x_0) \subset B_1,$$

then $u \in W^{2,p}_{\mathrm{loc}}(B_1)$. Moreover, there holds

$$\|u\|_{W^{2,p}(B_{1/2})} \leq C\{|u|_{L^\infty(B_1)} + \|f\|_{L^p(B_1)}\}$$

where C is a positive constant depending only on n, λ, Λ, and p.

As before we prove the following result instead.

THEOREM 5.23 *Suppose $u \in C(B_{8\sqrt{n}})$ is a viscosity solution of*

$$a_{ij}D_{ij}u = f \quad \text{in } B_{8\sqrt{n}}.$$

Then for any $p \in (n, \infty)$ there exist positive constants ε and C, depending only on n, λ, Λ, and p, such that if

$$\|u\|_{L^\infty(B_{8\sqrt{n}})} \leq 1, \quad \|f\|_{L^p(B_{8\sqrt{n}})} \leq \varepsilon,$$

and

$$\left(\frac{1}{|B_r(x_0)|}\int_{B_r(x_0)} |a_{ij} - a_{ij}(x_0)|^n\right)^{\frac{1}{n}} \leq \varepsilon \quad \text{for any } B_r(x_0) \subset B_{8\sqrt{n}},$$

then $u \in W^{2,p}(B_1)$ and $\|u\|_{W^{2,p}(B_1)} \leq C$.

Before the proof we first describe the strategy. Let Ω be a bounded domain and u be a continuous function in Ω. As in Section 2, we define for $M > 0$

$$\begin{aligned} G_M(u,\Omega) = \{x_0 \in \Omega : \ & \text{there exists an affine function } L \text{ such that} \\ & L(x) - \tfrac{M}{2}|x - x_0|^2 \leq u(x) \leq L(x) + \tfrac{M}{2}|x - x_0|^2 \\ & \text{for } x \in \Omega \text{ with equality at } x_0\}, \end{aligned}$$

$$A_M(u,\Omega) = \Omega \setminus G_M(u,\Omega).$$

We consider the function

$$\theta(x) = \theta(u,\Omega)(x) = \inf\{M : x \in G_M(u,\Omega)\} \in [0,\infty] \quad \text{for } x \in \Omega.$$

It is straightforward to verify that for $p \in (1, \infty]$ the condition $\theta \in L^p(\Omega)$ implies $D^2u \in L^p(\Omega)$ and

$$\|D^2u\|_{L^p(\Omega)} \leq 2\|\theta\|_{L^p(\Omega)}.$$

In order to study the integrability of the function θ we discuss its distribution function, that is,

$$\mu_\theta(t) = |\{x \in \Omega : \theta(x) > t\}| \quad \text{for any } t > 0.$$

It is clear that

$$\mu_\theta(t) \leq |A_t(u, \Omega)| \quad \text{for any } t > 0.$$

Hence we need to study the decay of $|A_t(u, \Omega)|$.

LEMMA 5.24 *Suppose that Ω is a bounded domain with $B_{8\sqrt{n}} \subset \Omega$ and that $u \in C(\Omega)$ is a viscosity solution of*

$$a_{ij}D_{ij}u = f \quad \text{in } B_{8\sqrt{n}}.$$

Then for any $\varepsilon_0 \in (0, 1)$ there exist an $M > 1$, depending only on n, λ, and Λ, and an $\varepsilon \in (0, 1)$, depending only on n, λ, Λ, and ε_0, such that if

$$\|f\|_{L^n(B_{8\sqrt{n}})} \leq \varepsilon, \quad \|a_{ij} - a_{ij}(0)\|_{L^n(B_{7\sqrt{n}})} \leq \varepsilon, \tag{5.21}$$

and

$$G_1(u, \Omega) \cap Q_3 \neq \varnothing, \tag{5.22}$$

then there holds

$$|G_M(u, \Omega) \cap Q_1| \geq 1 - \varepsilon_0.$$

PROOF: Let $x_1 \in G_1(u, \Omega) \cap Q_3$. Then there exists an affine function L such that

$$-\frac{1}{2}|x - x_1|^2 \leq u(x) - L(x) \leq \frac{1}{2}|x - x_1|^2 \quad \text{in } \Omega.$$

By considering $(u - L)/c(n)$ instead of u, for $c(n) > 1$ large enough, depending only on n, we may assume that

$$|u| \leq 1 \quad \text{in } B_{8\sqrt{n}}, \tag{5.23}$$

which implies

$$-|x|^2 \leq u(x) \leq |x|^2 \quad \text{for any } x \in \Omega \setminus B_{6\sqrt{n}}. \tag{5.24}$$

Solve for $h \in C(\bar{B}_{7\sqrt{n}}) \cap C^\infty(B_{7\sqrt{n}})$ such that

$$a_{ij}(0)D_{ij}h = 0 \quad \text{in } B_{7\sqrt{n}},$$
$$h = u \quad \text{on } \partial B_{7\sqrt{n}}.$$

Then Lemma 5.18 implies

$$|u - h|_{L^\infty(B_{6\sqrt{n}})} \leq C\{\varepsilon^\gamma + \|f\|_{L^n(B_{8\sqrt{n}})}\} \tag{5.25}$$

and

$$\|h\|_{C^2((B_{6\sqrt{n}})} \leq C \tag{5.26}$$

where $C > 0$ and $\gamma \in (0,1)$, as in Lemma 5.18, depend only on n, λ, and Λ. Consider $h|_{\bar{B}_{6\sqrt{n}}}$. Extend h outside $\bar{B}_{6\sqrt{n}}$ continuously such that $h = u$ in $\Omega \setminus B_{7\sqrt{n}}$ and $|u-h|_{L^\infty(\Omega)} = |u-h|_{L^\infty(B_{6\sqrt{n}})}$. Note $|h| \le 1$ in Ω. It follows that $|u-h|_{L^\infty(\Omega)} \le 2$ and hence with (5.24)

$$-2 - |x|^2 \le h(x) \le 2 + |x|^2 \qquad \text{for any } x \in \Omega \setminus \bar{B}_{6\sqrt{n}}.$$

Then there exists an $N > 1$, depending only on n, λ, and Λ, such that

$$Q_1 \subset G_N(h, \Omega). \tag{5.27}$$

Consider

$$w = \frac{\min\{1, \delta_0\}}{2C\varepsilon^\gamma}(u-h)$$

where δ_0 is the constant in Lemma 5.15 and C and γ are constants in (5.25) and (5.26). It is easy to check that w satisfies the hypotheses of Lemma 5.15 in Ω. We may apply Lemma 5.15 to get

$$|A_t(w, \Omega) \cap Q_1| \le Ct^{-\mu} \quad \text{for any } t > 0.$$

Therefore we have

$$|A_s(u-h, \Omega) \cap Q_1| \le C\varepsilon^{\gamma\mu}s^{-\mu} \quad \text{for any } s > 0.$$

It follows that

$$|G_N(u-h, \Omega) \cap Q_1| \ge 1 - C_1\varepsilon^{\gamma\mu} \ge 1 - \varepsilon_0$$

if we choose $\varepsilon = \varepsilon(n, \lambda, \Lambda, \varepsilon_0) \in (0,1)$ small. With (5.27) we get

$$|G_{2N}(u, \Omega) \cap Q_1| \ge 1 - \varepsilon_0.$$

□

REMARK 5.25. In fact, we prove Lemma 5.24 with assumption (5.22) replaced by (5.23).

PROOF OF THEOREM 5.23: Our proof has three steps.

Step 1. For any $\varepsilon_0 \in (0,1)$ there exist an $M > 1$, depending only on n, λ, and Λ, and an $\varepsilon \in (0,1)$, depending only on n, λ, Λ, and ε_0, such that under the assumptions of Theorem 5.23 there holds

$$|G_M(u, B_{8\sqrt{n}}) \cap Q_1| \ge 1 - \varepsilon_0. \tag{5.28}$$

We remark that M does not depend on ε_0. In fact, we have $|u| \le 1 \le |x|^2$ in $B_{8\sqrt{n}} \setminus B_{6\sqrt{n}}$. We may apply Lemma 5.24 to get (5.28) with $\Omega = B_{8\sqrt{n}}$ (see Remark 5.25).

Step 2. We set, for $k = 0, 1, \dots,$

$$A = A_{M^{k+1}}(u, B_{8\sqrt{n}}) \cap Q_1,$$

$$B = (A_{M^k}(u, B_{8\sqrt{n}}) \cap Q_1) \cup \{x \in Q_1 : m(f^n)(x) \ge (c_1M^k)^n\},$$

for some $c_1 > 0$ to be determined, depending only on n, λ, Λ, and ε_0. Then there holds

$$|A| \le \varepsilon_0 |B|.$$

The proof is identical to that of Lemma 5.15.

Step 3. We finish the proof of Lemma 5.24. We take ε_0 such that

$$\varepsilon_0 M^p = \frac{1}{2}$$

where M, depending only on n, λ, and Λ, is as in Step 1. Hence the constants ε and c_1 depend only on n, λ, Λ, and p. Define for $k = 0, 1, \dots,$

$$\alpha_k = |A_{M^k}(u, B_{8\sqrt{n}}) \cap Q_1|,$$
$$\beta_k = |\{x \in Q_1 : m(f^n)(x) \ge (c_1 M^k)^n\}|.$$

Then Step 2 implies $\alpha_{k+1} \le \varepsilon_0(\alpha_k + \beta_k)$ for any $k = 0, 1, \dots$. Hence by iteration we have

$$\alpha_k \le \varepsilon_0^k + \sum_{i=1}^{k-1} \varepsilon_0^{k-i} \beta_i.$$

Since $f^n \in L^{p/n}$ and the maximal operator is of strong type (p, p), we conclude that $m(f^n) \in L^{p/n}$ and

$$\|m(f^n)\|_{L^{p/n}} \le C\|f\|_{L^p}^n \le C.$$

Then the definition of β_k implies

$$\sum_{k \ge 0} M^{pk} \beta_k \le C.$$

As before we set

$$\theta(x) = \theta(u, B_{1/2})(x) = \inf\{M : x \in G_M(u, B_{1/2})\} \in [0, \infty] \quad \text{for } x \in B_{1/2}$$

and

$$\mu_\theta(t) = |\{x \in B_{1/2} : \theta(x) > t\}| \quad \text{for any } t > 0.$$

The proof will be finished if we show

$$\|\theta\|_{L^p(B_{1/2})} \le C.$$

It is clear that

$$\mu_\theta(t) \le |A_t(u, B_{1/2})| \le |A_t(u, B_{8\sqrt{n}}) \cap Q_1| \quad \text{for any } t > 0.$$

It suffices to prove, with the definition of α_k, that

$$\sum_{k \ge 1} M^{pk} \alpha_k \le C.$$

In fact, we have

$$\begin{aligned}\sum_{k\geq 1} M^{pk}\alpha_k &\leq \sum_{k\geq 1}(\varepsilon_0 M^p)^k + \sum_{k\geq 1}\sum_{i=0}^{k-1}\varepsilon_0^{k-i}M^{p(k-i)}M^{pi}\beta_i \\ &\leq \sum_{k\geq 1}2^{-k} + \Big(\sum_{i\geq 0}M^{pi}\beta_i\Big)\Big(\sum_{j\geq 1}2^{-j}\Big) \leq C.\end{aligned}$$

This finishes the proof. □

5.6. Global Estimates

In the previous two sections, we derived interior Schauder estimates and $W^{2,p}$-estimates for viscosity solutions. In fact, these estimates hold globally. In this section, we state these results without proof for classical solutions of Dirichlet problems for general linear elliptic equations. These results will be needed in the next chapter.

Let Ω be a bounded domain in $\mathbb{R}^n$, a_{ij} be continuous functions in Ω, and b_i and c be bounded functions in Ω. For some bounded function f in Ω and continuous function φ on $\partial\Omega$, consider

$$\begin{aligned} a_{ij}D_{ij}u + b_iD_iu + cu &= f \quad \text{in } \Omega, \\ u &= \varphi \quad \text{on } \partial\Omega. \end{aligned} \tag{5.29}$$

We always assume

$$a_{ij}(x)\xi_i\xi_j \geq \lambda|\xi|^2 \quad \text{for any } x\in\Omega \text{ and } \xi\in\mathbb{R}^n$$

for some constant $\lambda > 0$. In the following, we may require that φ be defined in Ω.

We first state the global Schauder estimate.

THEOREM 5.26 *For some constant $\alpha \in (0,1)$, let Ω be a bounded $C^{2,\alpha}$-domain in $\mathbb{R}^n$, and a_{ij}, b_i, and c be $C^\alpha(\bar\Omega)$-functions. Suppose $u \in C^{2,\alpha}(\bar\Omega)$ is a solution of* (5.29) *for some $f\in C^\alpha(\bar\Omega)$ and $\varphi \in C^{2,\alpha}(\bar\Omega)$. Then*

$$\|u\|_{C^{2,\alpha}(\bar\Omega)} \leq C\{\|u\|_{L^\infty(\Omega)} + \|f\|_{C^\alpha(\bar\Omega)} + \|\varphi\|_{C^{2,\alpha}(\bar\Omega)}\},$$

where C is a positive constant depending only on n, α, λ, Ω, and the $C^\alpha(\bar\Omega)$-norms of a_{ij}, b_i, and c.

Next, we state the global $W^{2,p}$-estimate.

THEOREM 5.27 *Let Ω be a bounded $C^{1,1}$-domain in $\mathbb{R}^n$, a_{ij} be continuous functions in Ω, and b_i and c be bounded functions in Ω. For some constant $p>1$, suppose $u \in W^{2,p}(\Omega)$ is a solution of* (5.29) *for some $f \in L^p(\Omega)$ and $\varphi\in W^{2,p}(\Omega)$. Then*

$$\|u\|_{W^{2,p}(\Omega)} \leq C\{\|u\|_{L^p(\Omega)} + \|f\|_{L^p(\Omega)} + \|\varphi\|_{W^{2,p}(\Omega)}\},$$

where C is a positive constant depending only on n, p, λ, Ω, the moduli of continuity of a_{ij}, and the $L^\infty(\Omega)$-norms of a_{ij}, b_i, and c.

By Sobolev embedding, we have the following result on $C^{1,\alpha}$-norms.

COROLLARY 5.28 *Let Ω be a bounded $C^{1,1}$-domain in $\mathbb{R}^n$, a_{ij} be continuous functions in Ω, and b_i, c be bounded functions in Ω. For some constant $p > n$, suppose $u \in W^{2,p}(\Omega)$ is a solution of (5.29) for some $f \in L^p(\Omega)$ and $\varphi \in W^{2,p}(\Omega)$. Then*

$$\|u\|_{C^{1,1-\frac{n}{p}}(\Omega)} \le C\{\|u\|_{L^p(\Omega)} + \|f\|_{L^p(\Omega)} + \|\varphi\|_{W^{2,p}(\Omega)}\},$$

where C is a positive constant depending only on n, p, λ, Ω, the moduli of continuity of a_{ij} and the $L^\infty(\Omega)$-norms of a_{ij}, b_i, and c.

We need to point out that Corollary 5.28 can be proved directly, without using Theorem 5.27.

CHAPTER 6

Existence of Solutions

In this chapter we will discuss the existence of solutions of some boundary value problems for elliptic differential equations. We will illustrate several methods.

6.1. Perron Method

In this section we will discuss the existence of solutions by the Perron method, where we prove the existence of solutions of Dirichlet problems for elliptic operators on general domains when solutions of the same problems on balls are known to exist. We will illustrate this by the Laplace operator.

Let Ω be a bounded domain in $\mathbb{R}^n$ and φ be a continuous function on $\partial\Omega$. Consider

$$\begin{aligned} \Delta u &= 0 \quad \text{in } \Omega, \\ u &= \varphi \quad \text{on } \partial\Omega. \end{aligned} \tag{6.1}$$

If Ω is a ball, then the solution of (6.1) is given by the Poisson integral formula. We now solve (6.1) by Perron's method. The maximum principle plays an essential role. In discussions below, we avoid mean value properties of harmonic functions.

We first define continuous subharmonic and subharmonic functions based on the maximum principle.

DEFINITION 6.1 Let Ω be a domain in $\mathbb{R}^n$ and v be a continuous function in Ω. Then v is *subharmonic* (*superharmonic*) in Ω if for any ball $B \subset \Omega$ and any harmonic function $w \in C(\bar{B})$

$$v \leq (\geq)\, w \text{ on } \partial B \quad \text{implies} \quad v \leq (\geq)\, w \text{ in } B.$$

We now prove a maximum principle for such subharmonic and superharmonic functions.

LEMMA 6.2 *Let Ω be a bounded domain in $\mathbb{R}^n$ and $u, v \in C(\bar{\Omega})$. Suppose u is subharmonic in Ω and v is a superharmonic in Ω with $u \leq v$ on $\partial\Omega$. Then $u \leq v$ in Ω.*

PROOF: Without loss of generality, we assume Ω is connected. We first note that $u - v \leq 0$ on $\partial\Omega$. Set $M = \max_{\bar{\Omega}}(u - v)$ and

$$D = \{x \in \Omega : u(x) - v(x) = M\} \subset \Omega.$$

It is obvious that D is relatively closed. This follows easily from the continuity of u and v.

Next we show that D is open. For any $x_0 \in D$, we take $r < \text{dist}(x_0, \partial\Omega)$. Let $\overline{u}$ and $\overline{v}$ solve, respectively,

$$\Delta\overline{u} = 0 \text{ in } B_r(x_0), \quad \overline{u} = u \text{ on } \partial B_r(x_0),$$
$$\Delta\overline{v} = 0 \text{ in } B_r(x_0), \quad \overline{v} = v \text{ on } \partial B_r(x_0).$$

The existence of $\overline{u}$ and $\overline{v}$ in $B_r(x_0)$ is implied by the Poisson integral formula. The definition of subsolutions and supersolutions implies $u \le \overline{u}$ and $\overline{v} \le v$ in $B_r(x_0)$; hence

$$\overline{u} - \overline{v} \ge u - v \quad \text{in } B_r(x_0).$$

Next,

$$\Delta(\overline{u} - \overline{v}) = 0 \qquad \text{in } B_r(x_0),$$
$$\overline{u} - \overline{v} = u - v \quad \text{on } \partial B_r(x_0).$$

With $u - v \le M$ on $\partial B_r(x_0)$, the maximum principle implies $\overline{u} - \overline{v} \le M$ in $B_r(x_0)$; in particular,

$$M \ge (\overline{u} - \overline{v})(x_0) \ge (u - v)(x_0) = M.$$

Hence $(\overline{u} - \overline{v})(x_0) = M$ and then $\overline{u} - \overline{v}$ has an interior maximum at x_0. By the strong maximum principle, $\overline{u} - \overline{v} \equiv M$ in $\overline{B}_r(x_0)$. Therefore, $u - v = M$ on $\partial B_r(x_0)$. This holds for any $r < \text{dist}(x_0, \partial\Omega)$. Then $u - v = M$ in $B_r(x_0)$ and hence $B_r(x_0) \subset D$. In conclusion, D is both relatively closed and open in Ω. Therefore either $D = \varnothing$ or $D = \Omega$. In other words, $u - v$ either attains its maximum only on $\partial\Omega$ or $u - v$ is constant. By $u \le v$ in $\partial\Omega$, we have $u \le v$ in Ω for both cases. □

The proof in fact yields the strong maximum principle: Either $u < v$ in Ω or $u - v$ is constant in Ω.

Before we start our discussion of Perron's method, we demonstrate how to generate bigger subharmonic functions from existing subharmonic functions.

LEMMA 6.3 *Let $v \in C(\overline{\Omega})$ be a subharmonic function in Ω and B be a ball in Ω with $\overline{B} \subset \Omega$. Let w be defined by $w = v$ in $\overline{\Omega} \setminus B$ and $\Delta w = 0$ in B. Then w is a subharmonic function in Ω and $v \le w$ in $\overline{\Omega}$.*

The function w is often called the *harmonic lifting* of v (in B).

PROOF: The existence of w in B is implied by the Poisson integral formula. Then w is smooth in B and is continuous in $\overline{\Omega}$. We also have $v \le w$ in B by Definition 6.1.

Next, we take any B' with $\overline{B}' \subset \Omega$ and consider a harmonic function $u \in C(\overline{B}')$ with $w \le u$ on $\partial B'$. By $v \le w$ on $\partial B'$, we have $v \le u$ on $\partial B'$. Then, v is subharmonic and u is harmonic in B' with $v \le u$ on $\partial B'$. By Lemma 6.2, we have $v \le u$ in B'. Hence $w \le u$ in $B \setminus B'$. Next, both w and u are harmonic in $B \cap B'$ and $w \le u$ on $\partial(B \cap B')$. By the maximum principle, we have $w \le u$ in $B \cap B'$. Hence $w \le u$ in B'. Therefore, w is subharmonic in Ω by Definition 6.1. □

Next, we solve (6.1) by the Perron method. Set

$$u_\varphi(x) = \sup\{v(x) : v \in C(\bar{\Omega}) \text{ is subharmonic in } \Omega \text{ and } v \le \varphi \text{ on } \partial\Omega\}. \tag{6.2}$$

In the first step in the Perron method, we prove that u_φ in (6.2) is a harmonic function in Ω.

LEMMA 6.4 *Let Ω be a bounded domain in $\mathbb{R}^n$ and φ be a continuous function on $\partial\Omega$. Then u_φ defined in (6.2) is harmonic in Ω.*

PROOF: Set

$$\mathcal{S}_\varphi = \{v : v \in C(\bar{\Omega}) \text{ is subharmonic in } \Omega \text{ and } v \le \varphi \text{ on } \partial\Omega\}.$$

Then for any $x \in \Omega$

$$u_\varphi(x) = \sup\{v(x) : v \in \mathcal{S}_\varphi\}.$$

In the following, we simply write $\mathcal{S} = \mathcal{S}_\varphi$.

Step 1. *We prove that u_φ is well defined.* To do this, we set

$$m = \min_{\partial\Omega} \varphi, \quad M = \max_{\partial\Omega} \varphi.$$

We note that the constant function m is in $\mathcal{S}$ and hence the set $\mathcal{S}$ is not empty. Next, the constant function M is obviously harmonic in Ω with $\varphi \le M$ on $\partial\Omega$. By Lemma 6.2, for any $v \in \mathcal{S}$

$$v \le M \quad \text{in } \bar{\Omega}.$$

Thus u_φ is well defined and $u_\varphi \le M$ in Ω.

Step 2. *We claim that $\mathcal{S}$ is closed by taking the maximum among finitely many functions in $\mathcal{S}$.* We take arbitrary $v_1, v_2, \ldots, v_k \in \mathcal{S}$ and set

$$v = \max\{v_1, v_2, \ldots, v_k\}.$$

It follows easily from Definition 6.1 that v is subharmonic in Ω. Hence $v \in \mathcal{S}$.

Step 3. *We prove that u_φ is harmonic in any $B_r(x_0) \subset \Omega$.* First, by the definition of u_φ, there exists a sequence of functions $v_i \in \mathcal{S}$ such that

$$\lim_{i\to\infty} v_i(x_0) = u_\varphi(x_0).$$

We may replace v_i above by any $\widetilde{v}_i \in \mathcal{S}$ with $\widetilde{v}_i \ge v_i$ since

$$v_i(x_0) \le \widetilde{v}_i(x_0) \le u_\varphi(x_0).$$

Replacing, if necessary, v_i by $\max\{m, v_i\} \in \mathcal{S}$, we may also assume

$$m \le v_i \le u_\varphi \quad \text{in } \Omega.$$

For the fixed $B_r(x_0)$ and each v_i, we let w_i be the harmonic lifting in Lemma 6.3. In other words, $w_i = v_i$ in $\Omega \setminus B_r(x_0)$ and

$$\begin{aligned} \Delta w_i &= 0 \quad \text{in } B_r(x_0), \\ w_i &= v_i \quad \text{on } \partial B_r(x_0). \end{aligned}$$

By Lemma 6.3, $w_i \in \mathcal{S}$ and $v_i \le w_i$ in Ω. Moreover, w_i is harmonic in $B_r(x_0)$ and satisfies

$$\lim_{i\to\infty} w_i(x_0) = u_\varphi(x_0),$$
$$m \le w_i \le u_\varphi \quad \text{in } \Omega,$$

for any $i = 1, 2, \ldots$. By the compactness of bounded harmonic functions, there exists a harmonic function w in $B_r(x_0)$ such that a subsequence of $\{w_i\}$, still denoted by $\{w_i\}$, converges to w in any compact subset of $B_r(x_0)$. We then easily conclude

$$w \le u_\varphi \text{ in } B_r(x_0) \quad \text{and} \quad w(x_0) = u_\varphi(x_0).$$

We now claim $u_\varphi = w$ in $B_r(x_0)$. To see this, we take any $\bar{x} \in B_r(x_0)$ and proceed similarly as before, with $\bar{x}$ replacing x_0. By the definition of u_φ, there exists a sequence of functions $\bar{v}_i \in \mathcal{S}$ such that

$$\lim_{i\to\infty} \bar{v}_i(\bar{x}) = u_\varphi(\bar{x}).$$

Replacing, if necessary, $\bar{v}_i$ by $\max\{\bar{v}_i, w_i\} \in \mathcal{S}$, we may also assume

$$w_i \le \bar{v}_i \le u_\varphi \quad \text{in } \Omega.$$

For the fixed $B_r(x_0)$ and each $\bar{v}_i$, we let $\bar{w}_i$ be the harmonic lifting in Lemma 6.3. Then, $\bar{w}_i \in \mathcal{S}$ and $\bar{v}_i \le \bar{w}_i$ in Ω. Moreover, $\bar{w}_i$ is harmonic in $B_r(x_0)$ and satisfies

$$\lim_{i\to\infty} \bar{w}_i(\bar{x}) = u_\varphi(\bar{x}),$$
$$m \le \max\{\bar{v}_i, w_i\} \le \bar{w}_i \le u_\varphi \quad \text{in } \Omega,$$

for any $i = 1, 2, \ldots$. By the compactness of bounded harmonic functions again, there exists a harmonic function $\bar{w}$ in $B_r(x_0)$ such that a subsequence of $\bar{w}_i$ converges to $\bar{w}$ in any compact subset of $B_r(x_0)$. We then easily conclude

$$w \le \bar{w} \le u_\varphi \qquad \text{in } B_r(x_0),$$
$$w(x_0) = \bar{w}(x_0) = u_\varphi(x_0),$$
$$\bar{w}(\bar{x}) = u_\varphi(\bar{x}).$$

We first note that $w - \bar{w}$ is a harmonic function in $B_r(x_0)$ with a maximum attained at x_0. By applying the strong maximum principle to $w - \bar{w}$ in $B_{r'}(x_0)$ for any $r' < r$, we conclude that $w - \bar{w}$ is constant, which is obviously 0. This implies $w = \bar{w}$ in $B_r(x_0)$, and in particular, $w(\bar{x}) = \bar{w}(\bar{x}) = u_\varphi(\bar{x})$. We then have $w = u_\varphi$ in $B_r(x_0)$ since $\bar{x}$ is arbitrary in $B_r(x_0)$. Therefore, u_φ is harmonic in $B_r(x_0)$. □

We note that u_φ in Lemma 6.4 is defined only in Ω. To discuss limits of $u_\varphi(x)$ as x approaches the boundary, we need to impose additional assumptions on the boundary $\partial\Omega$.

LEMMA 6.5 *Let φ be a continuous function on $\partial\Omega$ and u_φ be the function defined in (6.2). For some $x_0 \in \partial\Omega$, suppose $w_{x_0} \in C(\bar{\Omega})$ is a subharmonic function in Ω such that*

$$w_{x_0}(x_0) = 0, \quad w_{x_0}(x) < 0 \text{ for any } x \in \partial\Omega \setminus \{x_0\}; \tag{6.3}$$

then

$$\lim_{x \to x_0} u_\varphi(x) = \varphi(x_0).$$

PROOF: As in the proof of Lemma 6.4, we set

$$\mathcal{S}_\varphi = \{v : v \in C(\bar{\Omega}) \text{ is subharmonic in } \Omega \text{ and } v \le \varphi \text{ on } \partial\Omega\}.$$

We simply write $w = w_{x_0}$ and set $M = \max_{\partial\Omega} |\varphi|$.

Let ε be an arbitrary positive constant. By the continuity of φ at x_0, there exists a positive constant δ such that

$$|\varphi(x) - \varphi(x_0)| \le \varepsilon$$

for any $x \in \partial\Omega \cap B_\delta(x_0)$. We then choose K sufficiently large so that

$$-Kw(x) \ge 2M$$

for any $x \in \partial\Omega \setminus B_\delta(x_0)$; hence,

$$|\varphi - \varphi(x_0)| \le \varepsilon - Kw \quad \text{on } \partial\Omega.$$

Since $\varphi(x_0) - \varepsilon + Kw(x)$ is a subharmonic function in Ω with $\varphi(x_0) - \varepsilon + Kw \le \varphi$ on $\partial\Omega$, we have $\varphi(x_0) - \varepsilon + Kw \in \mathcal{S}_\varphi$. The definition of u_φ implies

$$\varphi(x_0) - \varepsilon + Kw \le u_\varphi \quad \text{in } \Omega.$$

On the other hand, $\varphi(x_0) + \varepsilon - Kw$ is a superharmonic in Ω with $\varphi(x_0) + \varepsilon - Kw \ge \varphi$ on Ω. Hence for any $v \in \mathcal{S}_\varphi$, we obtain by Lemma 6.2

$$v \le \varphi(x_0) + \varepsilon - Kw \quad \text{in } \Omega.$$

Again by the definition of u_φ, we have

$$u_\varphi \le \varphi(x_0) + \varepsilon - Kw \quad \text{in } \Omega.$$

Therefore,

$$|u_\varphi - \varphi(x_0)| \le \varepsilon - Kw \quad \text{in } \Omega,$$

which implies

$$\limsup_{x \to x_0} |u_\varphi(x) - \varphi(x_0)| \le \varepsilon.$$

We obtain the desired result by letting $\varepsilon \to 0$. □

The function w_{x_0} satisfying (6.3) is often called a *barrier function.* Barrier functions can be constructed for a large class of domains Ω. Take, for example, the case where Ω satisfies an *exterior sphere condition* at $x_0 \in \partial\Omega$ in the sense that there exists a ball $B_{r_0}(y_0)$ such that

$$\Omega \cap B_{r_0}(y_0) = \varnothing, \quad \bar{\Omega} \cap \bar{B}_{r_0}(y_0) = \{x_0\}.$$

To construct a barrier function at x_0, we set

$$w_{x_0}(x) = \Gamma(x - y_0) - \Gamma(x_0 - y_0) \quad \text{for any } x \in \bar{\Omega}$$

where Γ is the fundamental solution of the Laplace operator. It is easy to see that w_{x_0} is a harmonic function in Ω and satisfies (6.3). We note that the exterior sphere condition always holds for C^2-domains.

THEOREM 6.6 *Let Ω be a bounded domain in $\mathbb{R}^n$ satisfying the exterior sphere condition at every boundary point. Then for any $\varphi \in C(\partial\Omega)$, (6.1) admits a solution $u \in C^\infty(\Omega) \cap C(\bar{\Omega})$.*

6.2. Variational Method

In this section we discuss the Dirichlet problem for elliptic differential equations of divergence form and prove the existence of weak solutions.

Let Ω be a bounded domain in $\mathbb{R}^n$ and a_{ij}, b_i, and c be bounded functions in Ω. Consider the differential operator

$$Lu = -D_j(a_{ij}D_iu) + b_iD_iu + cu.$$

We always assume that

$$\lambda|\xi|^2 \le a_{ij}(x)\xi_i\xi_j \le \Lambda|\xi|^2$$

for any $x \in \Omega$ and $\xi \in \mathbb{R}^n$.

DEFINITION 6.7 Let $f \in L^2(\Omega)$ and $u \in H^1_{\mathrm{loc}}(\Omega)$. Then u is a weak solution of $Lu = f$ in Ω if

$$\int_\Omega (a_{ij}D_iuD_j\varphi + b_iD_iu\varphi + cu\varphi)dx = \int_\Omega f\varphi\,dx \tag{6.4}$$

for any $\varphi \in H^1_0(\Omega)$.

Next, we define

$$a(u, v) = \int_\Omega (a_{ij}D_iuD_jv + b_iD_iuv + cuv)dx$$

for any $u, v \in H^1_0(\Omega)$. We call a the bilinear form associated with the operator L. If $a_{ij} = a_{ji}$ and $c = 0$, then a is symmetric, i.e.,

$$a(u, v) = a(v, u) \quad \text{for any } u, v \in H^1_0(\Omega).$$

We now solve the Dirichlet problem in the weak sense for a special class of elliptic operators. We recall that the standard $H^1_0(\Omega)$ inner product is defined by

$$(u, v)_{H^1_0(\Omega)} = \int_\Omega \nabla u \cdot \nabla v\,dx.$$

THEOREM 6.8 *Let a_{ij}, b_i, and c be bounded functions in Ω and $f \in L^2(\Omega)$. Assume the bilinear form a associated with L is coercive; i.e.,*

$$a(u, u) \ge c_0\|u\|^2_{H^1_0(\Omega)}$$

for any $u \in H^1_0(\Omega)$. Then there exists a unique weak solution $u \in H^1_0(\Omega)$ of $Lu = f$.

PROOF: We define a linear functional F on $H_0^1(\Omega)$ by

$$F(\varphi) = \int_\Omega f\varphi\,dx$$

for any $\varphi \in H_0^1(\Omega)$. By the Cauchy inequality and the Poincarè inequality, we have

$$|F(\varphi)| \le \|f\|_{L^2(\Omega)}\|\varphi\|_{L^2(\Omega)} \le C\|f\|_{L^2(\Omega)}\|\varphi\|_{H_0^1(\Omega)}.$$

Hence F is a bounded linear functional on $H_0^1(\Omega)$.

We first consider a special case where a is symmetric. (This includes the case $a_{ij} = a_{ji}$ and $c \ge 0$.) It is easy to see that $a(u, v)$ is an inner product in $H_0^1(\Omega)$ that is equivalent to the standard $H_0^1(\Omega)$ inner product. By the Riesz representation theorem, there exists a $u \in H_0^1(\Omega)$ such that

$$a(u, \varphi) = F(\varphi)$$

for any $\varphi \in H_0^1(\Omega)$. Therefore, u is the desired solution.

We now consider the general case where a is not necessarily symmetric. We first note that

$$|a(u, v)| \le C\|u\|_{H_0^1(\Omega)}\|v\|_{H_0^1(\Omega)}$$

for any $u, v \in H_0^1(\Omega)$.

For each fixed $u \in H_0^1(\Omega)$, the mapping $v \mapsto a(u, v)$ is a bounded linear functional on $H_0^1(\Omega)$. By the Riesz representation theorem, there exists a unique $w \in H_0^1(\Omega)$ such that

$$a(u, v) = (w, v)_{H_0^1(\Omega)}$$

for any $v \in H_0^1(\Omega)$. Now we write $w = Au$; i.e.,

$$a(u, v) = (Au, v)_{H_0^1(\Omega)}$$

for any $u, v \in H_0^1(\Omega)$. It is straightforward to check that A is a linear operator on $H_0^1(\Omega)$.

Next,

$$\|Au\|^2_{H_0^1(\Omega)} = (Au, Au)_{H_0^1(\Omega)} = a(u, Au) \le C\|u\|_{H_0^1(\Omega)}\|Au\|_{H_0^1(\Omega)}$$

and hence

$$\|Au\|_{H_0^1(\Omega)} \le C\|u\|_{H_0^1(\Omega)} \quad \text{for any } u \in H_0^1(\Omega).$$

Therefore, $A : H_0^1(\Omega) \to H_0^1(\Omega)$ is a bounded linear operator. By the coerciveness, we have

$$c_0\|u\|^2_{H_0^1(\Omega)} \le a(u, u) = (Au, u)_{H_0^1(\Omega)} \le \|u\|_{H_0^1(\Omega)}\|Au\|_{H_0^1(\Omega)}$$

and hence

$$c_0\|u\|_{H_0^1(\Omega)} \le \|Au\|_{H_0^1(\Omega)}$$

for any $u \in H_0^1(\Omega)$. It follows that A is one-to-one and the range $R(A)$ of A is closed in $H_0^1(\Omega)$.

Next, for any $w \in R(A)^\perp$, we have

$$c_0 \|w\|^2_{H_0^1(\Omega)} \le a(w,w) = (Aw,w)_{H_0^1(\Omega)} = 0$$

and hence $w = 0$. Then $R(A)^\perp = \{0\}$ and hence $R(A) = H_0^1(\Omega)$; in other words, A is onto.

For the bounded linear functional F in $H_0^1(\Omega)$ introduced at the beginning of the proof, by the Riesz representation theorem, there exists a $w \in H_0^1(\Omega)$ such that

$$(w,v)_{H_0^1(\Omega)} = F(v)$$

for any $v \in H_0^1(\Omega)$. Since A is onto, we can find a $u \in H_0^1(\Omega)$ such that $Au = w$; hence,

$$a(u,v) = (Au,v)_{H_0^1(\Omega)} = (w,v)_{H_0^1(\Omega)} = F(v)$$

for any $v \in H_0^1(\Omega)$. This proves the existence.

For the uniqueness, we assume $\widetilde{u} \in H_0^1(\Omega)$ also satisfies

$$a(\widetilde{u},v) = F(v)$$

for any $v \in H_0^1(\Omega)$. Then

$$a(u-\widetilde{u},v) = 0$$

for any $v \in H_0^1(\Omega)$. With $v = u - \widetilde{u}$, we obtain

$$a(u-\widetilde{u},u-\widetilde{u}) = 0.$$

By the coerciveness, we have $u = \widetilde{u}$. □

We point out that the method in the proof of the general case, when formulated in an abstract form, is known as the Lax-Milgram theorem.

The Dirichlet problem can be solved for a larger class of elliptic equations of divergence form. However, the method is much more involved. In the following, we state a simple consequence of Theorem 6.8.

THEOREM 6.9 *Let a_{ij}, b_i, and c be bounded functions in Ω and $f \in L^2(\Omega)$. Then there exists a μ_0 depending only on a_{ij}, b_i, and c such that for any $\mu \ge \mu_0$ there exists a unique weak solution $u \in H_0^1(\Omega)$ of $(L+\mu)u = f$.*

PROOF: Define

$$a_\mu(u,v) = a(u,v) + (u,v)_{L^2(\Omega)}.$$

In other words, a_μ is the bilinear form associated with the operator $L+\mu$. It is easy to check that a_μ is coercive for μ sufficiently large. Then we may apply Theorem 6.8 to $L+\mu$. □

In the rest of this section, we use a minimizing process to solve the Dirichlet problem on the bounded domain with the homogeneous boundary value.

Let Ω be a bounded domain a_{ij}, and c be bounded functions in Ω satisfying

$$\lambda|\xi|^2 \le a_{ij}(x)\xi_i\xi_j \le \Lambda|\xi|^2$$

for any $x \in \Omega$ and $\xi \in \mathbb{R}^n$. Suppose $f \in L^2(\Omega)$. Define

$$J(u) = \frac{1}{2}\int_\Omega (a_{ij} D_i u D_j u + cu^2)dx + \int_\Omega uf\,dx. \tag{6.5}$$

THEOREM 6.10 *Let a_{ij} and c be bounded functions in Ω with $a_{ij} = a_{ji}$ and $c \geq 0$, and $f \in L^2(\Omega)$. Then J admits a minimizer $u \in H_0^1(\Omega)$.*

It is easy to check that the minimizer u is a weak solution of

$$-D_j(a_{ij} D_i u) + cu = f \quad \text{in } \Omega.$$

PROOF: We first prove that J has a lower bound in $H_0^1(\Omega)$. By the Poincarè inequality, we have for any $u \in \mathcal{H}_0^1(\Omega)$

$$\int_\Omega u^2\,dx \leq C\int_\Omega |\nabla u|^2\,dx,$$

where C is a positive constant depending only on Ω. Then

$$\begin{aligned}\int_\Omega |uf|dx &\leq \Big(\int_\Omega u^2\,dx\Big)^{\frac{1}{2}}\Big(\int_\Omega f^2\,dx\Big)^{\frac{1}{2}} \\ &\leq \sqrt{C}\Big(\int_\Omega |\nabla u|^2\,dx\Big)^{\frac{1}{2}}\Big(\int_\Omega f^2\,dx\Big)^{\frac{1}{2}} \\ &\leq \frac{1}{4\lambda}\int_\Omega |\nabla u|^2\,dx + C\lambda\int_\Omega f^2\,dx.\end{aligned}$$

Hence for any $u \in H_0^1(\Omega)$,

$$J(u) \geq \frac{1}{4\lambda}\int_\Omega |\nabla u|^2\,dx - C\lambda\int_\Omega f^2\,dx \tag{6.6}$$

and in particular

$$J(u) \geq -C\int_\Omega f^2\,dx.$$

Therefore, J has a lower bound in $H_0^1(\Omega)$. We set

$$J_0 = \inf\{J(u) : u \in \mathcal{H}_0^1(\Omega)\}.$$

Next, we prove that J_0 is attained by some $u \in H_0^1(\Omega)$. We consider a minimizing sequence $\{u_k\} \subset H_0^1(\Omega)$ with $J(u_k) \to J_0$ as $k \to \infty$. By (6.6) we have

$$\int_\Omega |\nabla u_k|^2\,dx \leq 4\lambda J(u_k) + 4C\lambda^2\int_y \Omega f^2\,dx.$$

The convergence of $J(u_k)$ obviously implies that $\{J(u_k)\}$ is a bounded sequence; then $\|u_k\|_{H_0^1}$ is uniformly bounded. By Rellich's theorem, there exists a subsequence $\{u_{k'}\}$ and a $u \in H_0^1(\Omega)$ such that

$$u_{k'} \to u \quad \text{in the } L^2\text{-norm as } k' \to \infty.$$

Next, with the help of the Hilbert space structure of $H_0^1(\Omega)$ it is not difficult to prove

$$J(u) \leq \liminf_{k' \to \infty} J(u'_k).$$

This implies $J(u) = J_0$. We conclude that J_0 is attained in $H_0^1(\Omega)$. □

We point out that the minimizing process is an important method in the calculus of variation.

6.3. Continuity Method

In this section we discuss how to solve Dirichlet problems by the method of continuity. We illustrate this method by solving the Dirichlet problem for uniformly elliptic equations on $C^{2,\alpha}$-domains by assuming that a similar problem for the Laplace equation can be solved. The method of continuity can be applied to nonlinear elliptic equations. The crucial ingredient is the a priori estimates.

Let Ω be a bounded domain in $\mathbb{R}^n$, and let a_{ij}, b_i, and c be defined in Ω, with $a_{ij} = a_{ji}$. We consider the operator L given by

$$Lu = a_{ij} D_{ij} u + b_i D_i u + cu \quad \text{in } \Omega \tag{6.7}$$

for any $u \in C^2(\Omega)$. The operator L is always assumed to be *uniformly elliptic* in Ω; namely,

$$a_{ij} \xi_i \xi_j \geq \lambda |\xi|^2 \quad \text{for any } x \in \Omega \text{ and } \xi \in \mathbb{R}^n \tag{6.8}$$

for some positive constant λ.

Now we state a general existence result for solutions of the Dirichlet problem with $C^{2,\alpha}$ boundary values for general uniformly elliptic equations with C^α coefficients.

THEOREM 6.11 *Let Ω be a bounded $C^{2,\alpha}$-domain in $\mathbb{R}^n$ and L be a uniformly elliptic operator in Ω as defined in* (6.7), *with* (6.8) *satisfied, $c \leq 0$ in Ω, and $a_{ij}, b_i, c \in C^\alpha(\bar{\Omega})$ for some $\alpha \in (0,1)$. Then for any $f \in C^\alpha(\bar{\Omega})$ and $\varphi \in C^{2,\alpha}(\bar{\Omega})$, there exists a* (*unique*) *solution $u \in C^{2,\alpha}(\bar{\Omega})$ of the Dirichlet problem*

$$Lu = f \quad \text{in } \Omega,$$
$$u = \varphi \quad \text{on } \partial\Omega.$$

Theorem 6.11 plays an extremely important role in the theory of elliptic differential equations of the second order. The crucial step in solving the Dirichlet problem for L is to assume that the similar Dirichlet problem for the Laplace operator is solved. Specifically, we prove the following result.

THEOREM 6.12 *Let Ω be a bounded $C^{2,\alpha}$-domain in $\mathbb{R}^n$ and L be a uniformly elliptic operator in Ω as defined in* (6.7)*, with* (6.8) *satisfied, $c \le 0$ in Ω, and $a_{ij}, b_i, c \in C^\alpha(\bar{\Omega})$ for some $\alpha \in (0,1)$. If the Dirichlet problem for the Poisson equation*

$$\Delta u = f \quad \text{in } \Omega,$$
$$u = \varphi \quad \text{on } \partial\Omega,$$

has a $C^{2,\alpha}(\bar{\Omega})$ solution for all $f \in C^\alpha(\bar{\Omega})$ and $\varphi \in C^{2,\alpha}(\bar{\Omega})$, then the Dirichlet problem

$$Lu = f \quad \text{in } \Omega,$$
$$u = \varphi \quad \text{on } \partial\Omega,$$

also has a (unique) $C^{2,\alpha}(\bar{\Omega})$ solution for all such f and φ.

The proof is based on the *method of continuity*. Briefly summarized, this method as applied here starts with the solution of the Poisson equation $\Delta u = f$ and then arrives at a solution of $Lu = f$ through solutions of a continuous family of equations connecting $\Delta u = f$ and $Lu = f$. The global $C^{2,\alpha}$-estimates play essential roles.

PROOF: Without loss of generality, we assume $\varphi = 0$. Otherwise, we consider $Lv = f - L\varphi$ in Ω, $v = 0$ on $\partial\Omega$.

We consider the family of equations

$$L_t u \equiv tLu + (1-t)\Delta u = f$$

for $t \in [0,1]$. We note that $L_0 = \Delta$, $L_1 = L$. By writing

$$L_t u = a_{ij}^t D_{ij} u + b_i^t D_i u + c^t u,$$

it is easy to see that

$$a_{ij}^t(x)\xi_i\xi_j \ge \min(1,\lambda)|\xi|^2$$

for any $x \in \Omega$ and $\xi \in \mathbb{R}^n$ and that

$$|a_{ij}^t|_{C^\alpha(\bar{\Omega})}, |b_i^t|_{C^\alpha(\bar{\Omega})}, |c^t|_{C^\alpha(\bar{\Omega})} \le \max(1,\Lambda)$$

independently of $t \in [0,1]$. It follows that

$$|L_t u|_{C^\alpha(\bar{\Omega})} \le C|u|_{C^{2,\alpha}(\Omega)}$$

where C is a positive constant depending only on n, α, λ, Λ, and Ω. Then for each $t \in [0,1]$, $L_t : \mathcal{X} \to C^\alpha(\Omega)$ is a bounded linear operator, where

$$\mathcal{X} = \{u \in C^{2,\alpha}(\bar{\Omega}) : u = 0 \text{ on } \partial\Omega\}.$$

We note that $\mathcal{X}$ is a Banach space with respect to $|\cdot|_{C^{2,\alpha}(\bar{\Omega})}$.

We now let I be the collection of $s \in [0,1]$ such that the Dirichlet problem

$$L_s u = f \quad \text{in } \Omega,$$
$$u = 0 \quad \text{on } \partial\Omega,$$

is solvable in $C^{2,\alpha}(\bar{\Omega})$ for any $f \in C^{\alpha}(\bar{\Omega})$. We take an $s \in I$ and let $u = L_s^{-1} f$ be the (unique) solution. The global $C^{2,\alpha}$-estimates and the maximum principle imply

$$|L_s^{-1} f|_{C^{2,\alpha}(\Omega)} \le C|f|_{C^{\alpha}(\bar{\Omega})}.$$

For any $t \in [0,1]$ and $f \in C^{\alpha}(\bar{\Omega})$, we write $L_t u = f$ as

$$L_s u = f + (L_s - L_t)u = f + (t-s)(\Delta u - Lu).$$

Hence $u \in C^{2,\alpha}(\bar{\Omega})$ is a solution of

$$\begin{aligned} L_t u &= f \quad \text{in } \Omega, \\ u &= 0 \quad \text{on } \partial\Omega, \end{aligned}$$

if and only if

$$u = L_s^{-1}(f + (t-s)(\Delta u - Lu)).$$

For any $u \in \mathcal{X}$, set

$$Tu = L_s^{-1}(f + (t-s)(\Delta u - Lu)).$$

Then $T : \mathcal{X} \to \mathcal{X}$ is an operator and for any $u, v \in \mathcal{X}$

$$\begin{aligned} |Tu - Tv|_{C^{2,\alpha}(\bar{\Omega})} &= \left|(t-s)L_s^{-1}((\Delta - L)(u-v))\right|_{C^{2,\alpha}(\bar{\Omega})} \\ &\le C|t-s||(\Delta - L)(u-v)|_{C^{\alpha}(\bar{\Omega})} \\ &\le C|t-s||u-v|_{C^{2,\alpha}(\bar{\Omega})}. \end{aligned}$$

Therefore, $T : \mathcal{X} \to \mathcal{X}$ is a contraction if

$$|t-s| < \delta = C^{-1}.$$

Hence for any $t \in [0,1]$ with $|t-s| < \delta$, there exists a $u \in \mathcal{X}$ such that $u = Tu$; i.e.,

$$u = L_s^{-1}(f + (t-s)(\Delta u - Lu)).$$

In other words, for any $t \in [0,1]$ with $|t-s| < \delta$ and any $f \in C^{\alpha}(\bar{\Omega})$, there exists a solution $u \in C^{2,\alpha}(\bar{\Omega})$ of

$$\begin{aligned} L_t u &= f \quad \text{in } \Omega, \\ u &= 0 \quad \text{on } \partial\Omega. \end{aligned}$$

Thus if $s \in I$, then $t \in I$ for any $t \in [0,1]$ with $|t-s| < \delta$. We now divide the interval $[0,1]$ into subintervals of length less than δ. By $0 \in I$, we conclude $1 \in I$. □

6.4. Compactness Methods

In this section we discuss several methods to solve nonlinear elliptic differential equations. All these methods involve the compactness of the Hölder functions: A bounded sequence of Hölder functions has a subsequence convergent to a Hölder function.

We first consider a class of semilinear elliptic equations.

THEOREM 6.13 *Let Ω be a bounded $C^{2,\alpha}$-domain in $\mathbb{R}^n$ and f be a C^1-function in $\bar{\Omega}\times\mathbb{R}$. Suppose $\underline{u},\bar{u}\in C^{2,\alpha}(\bar{\Omega})$ satisfy $\underline{u}\le\bar{u}$,*

$$\Delta\underline{u}\ge f(x,\underline{u})\ \text{in}\ \Omega,\quad \underline{u}\le 0\ \text{on}\ \partial\Omega,$$
$$\Delta\bar{u}\le f(x,\bar{u})\ \text{in}\ \Omega,\quad \bar{u}\ge 0\ \text{on}\ \partial\Omega.$$

Then there exists a solution $u\in C^{2,\alpha}(\bar{\Omega})$ of

$$\Delta u = f(x,u)\ \text{in}\ \Omega,\qquad u=0\ \text{on}\ \partial\Omega,\ \underline{u}\le u\le\bar{u}\ \text{in}\ \Omega.$$

We note that $\underline{u}$ and $\bar{u}$ are subsolutions and supersolutions, respectively.

PROOF: Set

$$m=\inf_{\Omega}\underline{u},\qquad M=\sup_{\Omega}\bar{u}.$$

We take $\lambda>0$ large so that

$$\lambda > f_z(x,z)\quad \text{for any } (x,z)\in\bar{\Omega}\times[m,M].$$

Now we write $u_0=\underline{u}$. For any u_k, $k=0,1,\dots$, we suppose $u_{k+1}\in C^{2,\alpha}(\bar{\Omega})$ solve

$$\begin{aligned}\Delta u_{k+1}-\lambda u_{k+1} &= f(x,u_k)-\lambda u_k && \text{in } \Omega,\\ u_{k+1} &= 0 && \text{on } \partial\Omega.\end{aligned}\tag{6.9}$$

We first prove

$$\underline{u}\le u_k\le\bar{u}\quad \text{in}\ \Omega.$$

This is obviously true for $k=0$. Suppose it holds for some $k\ge 0$. We now consider u_{k+1}. First, we note

$$\Delta(u_{k+1}-\underline{u})-\lambda(u_{k+1}-\underline{u})\le (f(x,u_k)-f(x,\underline{u}))-\lambda(u_k-\underline{u}).$$

By the mean value theorem, we have

$$(f(x,u_k)-f(x,\underline{u}))-\lambda(u_k-\underline{u})=-(\lambda-f_z(x,\theta))(u_k-\underline{u})$$

where θ is between $u_k(x)$ and $\underline{u}(x)$. Therefore, we obtain

$$\begin{aligned}\Delta(u_{k+1}-\underline{u})-\lambda(u_{k+1}-\underline{u})&\le 0 \quad \text{in } \Omega,\\ u_{k+1}-\underline{u}&\ge 0\quad \text{on } \partial\Omega.\end{aligned}$$

By the maximum principle, we have $u_{k+1}\ge\underline{u}$ in Ω. Similarly, we have $u_{k+1}\le\bar{u}$ in Ω. In particular, we have

$$m\le u_k(x)\le M\quad \text{for any } x\in\Omega \text{ and } k=0,1,\dots.$$

Similarly, we can prove

$$\underline{u}\le u_1\le u_2\le\cdots\le\bar{u}.$$

In other words, $\{u_k\}$ is an increasing sequence. Therefore, there exists a function u in Ω such that $u_k(x)\to u(x)$ as $k\to\infty$ for each $x\in\Omega$.

Next, the right-hand side expression in (6.9) is uniformly bounded independent of k. By the global $C^{1,\alpha}$-estimate, we have

$$\|u_k\|_{C^{1,\alpha}(\bar{\Omega})}\le C$$

where C is a positive constant depending only on n, λ, m, M, and Ω, independently of k. In particular, the right-hand side expression in (6.9) is uniformly bounded in $C^{1,\alpha}$-norms independent of k. By the global $C^{2,\alpha}$-estimate, we have

$$\|u_k\|_{C^{2,\alpha}(\bar{\Omega})} \leq C$$

where C is a positive constant depending only on n, λ, m, M, and Ω, independently of k. Therefore, $u \in C^2(\Omega)$ and

$$u_k \to u \quad \text{in the } C^2\text{-norm in } \Omega.$$

Hence u is the desired solution. □

As an application, we prove the following result.

COROLLARY 6.14 *Let Ω be a bounded $C^{2,\alpha}$-domain in $\mathbb{R}^n$ and f be a bounded C^1-function in $\bar{\Omega} \times \mathbb{R}$. Then there exists a solution $u \in C^{2,\alpha}(\bar{\Omega})$ of*

$$\begin{aligned} \Delta u &= f(x,u) \quad \text{in } \Omega, \\ u &= 0 \qquad\qquad \text{on } \partial\Omega. \end{aligned}$$

PROOF: Set

$$M = \sup_{\Omega\times\mathbb{R}} |f|.$$

Let $\underline{u}, \bar{u} \in C^{2,\alpha}(\bar{\Omega})$ satisfy

$$\begin{aligned} \Delta \underline{u} &= M \quad \text{in } \Omega, \quad \underline{u} = 0 \text{ on } \partial\Omega, \\ \Delta \bar{u} &= -M \text{ in } \Omega, \quad \bar{u} = 0 \text{ on } \partial\Omega; \end{aligned}$$

then

$$\Delta \underline{u} \geq \Delta \bar{u} \quad \text{in } \Omega, \quad \underline{u} = \bar{u} \text{ on } \partial\Omega.$$

By the maximum principle, we have $\underline{u} \leq \bar{u}$ in Ω. It is obvious that

$$\Delta \underline{u} \geq f(x, \underline{u}), \quad \Delta \bar{u} \leq f(x, \bar{u}) \text{ in } \Omega.$$

Hence $\underline{u}$ and $\bar{u}$ satisfy the conditions in Theorem 6.13. We obtain the desired result by Theorem 6.13. □

REMARK 6.15. Corollary 6.14 still holds if we assume f is C^1 in $\bar{\Omega} \times \mathbb{R}$ and satisfies

$$|f(x,z)| \leq C(1 + |z|^\tau) \quad \text{for any } (x,z) \in \bar{\Omega} \times \mathbb{R}$$

for some $C > 0$ and $\tau \in [0, 1)$. It is important to assume $\tau < 1$. Dirichlet problems may not be solvable if $\tau = 1$.

6.5. Single- and Double-Layer Potentials Methods

We begin with the Dirichlet problem for a half-space:

$$\begin{cases} \Delta u = 0 & \text{in } \mathbb{R}^{n+1}_+ = \{x \in \mathbb{R}^{n+1} : x_{n+1} > 0\}, \\ u = f & \text{on } \partial\mathbb{R}^{n+1}_+ = \mathbb{R}^n \times \{\underline{0}\}. \end{cases} \tag{6.10}$$

Using the Poisson integral formula, we can represent a solution as

$$u(x, y) = P_y * f(x), \quad (x, y) \in \mathbb{R}^{n+1}_+ = \mathbb{R}^n \times \mathbb{R}_+, \tag{6.11}$$

where

$$P_y(x) = \frac{\Gamma(\frac{n+1}{2})}{\pi^{\frac{n+1}{2}}} \frac{y}{(|x|^2 + |y|^2)^{\frac{n+1}{2}}}, \quad f \in C_0(\mathbb{R}^{n+1}).$$

The estimates for convolutions imply that

$$\sup_{y>0} \|u(\cdot, y)\|_{L^p(\mathbb{R}^n)} \leq \|f\|_{L^p(\mathbb{R}^n)} \quad \text{for all } 1 \leq p \leq \infty. \tag{6.12}$$

(Note that for $p = +\infty$, (6.12) is also a consequence of the maximum principle; and for $p = 1$, $\|u(\cdot, y)\|_{L^1(\mathbb{R}^n)} \leq \|P_y\|_{L^1(\mathbb{R}^n)} \cdot \|f\|_{L^1(\mathbb{R}^n)} = \|f\|_{L^1(\mathbb{R}^n)}$.)

We can also reverse the implication of (6.12) in the following sense (via Fatou's theorem): if a harmonic function u in $\mathbb{R}^{n+1}_+$ satisfies (6.12), then u has a nontangential limit a.e. on $\partial\mathbb{R}^{n+1}_+$, and the limit function $u_0 = u(\cdot, 0) \in L^p(\mathbb{R}^n)$ (if $p > 1$; if $p = 1$, then u_0 is a Radon measure) with $u(x, y) = P_y * u_0(x)$.

SKETCH OF PROOF: (See **[15]** for details.) Suppose u is harmonic in $\mathbb{R}^{n+1}_+$ with

$$\sup_{y>0} \|u(\cdot, y)\|_{L^p(\mathbb{R}^n)} < \infty. \tag{6.13}$$

Note $u(x, y+\rho) = P_y * u_\rho(x)$ where $u_\rho(x) = u(x, \rho))$, $y > 0$, $\rho > 0$. Statement (6.13) implies that $u_{\rho_n} \rightharpoonup v$ in $L^p(\mathbb{R}^n)$ for a sequence of $\rho_n \downarrow 0$. It is then easy to see that $P_y * u_{\rho_n}(x) \to P_y * v(x)$ for all $y > 0$ as $\rho_n \to 0^+$.

On the other hand, $P_y * u_{\rho_n}(x) = u(x, y + \rho_n)$. Thus $P_y * v(x) = u(x, y)$ where $v \in L^p(\mathbb{R}^n)$, and when $p = 1$ we naturally replace v by a Radon measure. □

Now we assume that Ω is a bounded, connected domain in $\mathbb{R}^n$, $n \geq 3$, with a C^2-boundary. (Here we assume $n \neq 2$ to simplify matters and avoid technicalities.) Consider the Dirichlet problem

$$\begin{cases} \Delta u = 0 & \text{in } \Omega, \\ u|_{\partial\Omega} = f & \in C(\partial\Omega). \end{cases} \tag{6.14}$$

Let $\gamma(x) = \frac{C_n}{|x|^{n-2}}$ be the fundamental solution of the Laplace operator in $\mathbb{R}^n$; here

$$C_n = -\frac{1}{(n-2)\omega_n} = \frac{-1}{(n-2)} \frac{\Gamma(\frac{n}{2})}{2\pi^{n/2}}.$$

Set $R(x, y) = -\gamma(x - y)$, and for $f \in C(\partial\Omega)$, we define the *double-layer potential*

$$\mathcal{D}f(p) = \int_{\partial\Omega} \frac{\partial}{\partial n_Q} R(P, Q) f(Q) d\mathcal{H}^{n-1}(Q), \quad P \notin \partial\Omega, \tag{6.15}$$

and the *single-layer potential*

$$S(f)(P) = \int_{\partial\Omega} R(P, Q) f(Q) d\mathcal{H}^{n-1}(Q), \quad P \notin \partial\Omega. \tag{6.16}$$

Here n_Q is the outward unit normal for $\partial\Omega$ at Q.

It is easy to check that

$$\Delta\mathcal{D}f(P) = 0 \quad \text{for } P \in \mathbb{R}^n \mid \partial\Omega.$$

We need to understand the boundary behavior of $\mathcal{D}f(P)$ on $\partial\Omega$.

LEMMA 6.16 *If $f \in C(\partial\Omega)$, then*

(i) *$\mathcal{D}f \in C(\overline{\Omega})$ and*

(ii) *$\mathcal{D}f \in C(\overline{\Omega^c})$.*

In other words, $\mathcal{D}f$ can be extended continuously from inside Ω to $\overline{\Omega}$, and from outside Ω to $\overline{\Omega^c}$. Let $\mathcal{D}_+ f$ and $\mathcal{D}_- f$ be the restrictions of these two functions to $\partial\Omega$. Set

$$K(P, Q) = \frac{\partial}{\partial n_Q} R(P, Q) = \frac{1}{\omega_n} \frac{\langle P - Q, n_Q\rangle}{|P - Q|^n}.$$

Thus

$$K \in C(\partial\Omega \times \partial\Omega \setminus \{(P, P) : P \in \partial\Omega\})$$

and $|K(P, Q)| \leq C/|P - Q|^{n-2}$ for $P, Q \in \partial\Omega$ and some $C < \infty$. The latter estimate follows from the C^2 property of $\partial\Omega$. We shall define, for $f \in C(\partial\Omega)$, the operator

$$Tf(P) = \int_{\partial\Omega} K(P, Q) f(Q) d\mathcal{H}^{n-1}(Q), \quad P \in \partial\Omega. \tag{6.17}$$

We have the following:

LEMMA 6.17 (Jump Relations for $\mathcal{D}$)

(i) *$\mathcal{D}_+ = \frac{1}{2} I + T$ and*

(ii) *$\mathcal{D}_- = -\frac{1}{2} I + T$.*

Moreover, $T : C(\partial\Omega) \to C(\partial\Omega)$ is compact.

PROOF OF LEMMAS 6.16 AND 6.17: We first verify that T defined by (6.17) is a compact operator from $C(\partial\Omega) \to C(\partial\Omega)$. Let

$$K_N(P, Q) = \operatorname{sign} K(P, Q) \cdot \min\{N, |K(P, Q)|\}, \quad N \in \mathbb{Z}_+.$$

Thus K_N is continuous on $\partial\Omega \times \partial\Omega$, and the Arzela-Ascoli theorem implies that $T_N f = \int_{\partial\Omega} K_N(P, Q) f(Q) d\mathcal{H}^{n-1}(Q)$ is compact on $C(\partial\Omega)$. Furthermore, since $\|T_N\| \leq \sup_{P\in\partial\Omega} \|K_N(P,Q)\|_{L^1(\partial\Omega)} \leq C < \infty$ where C is independent of N, it is rather easy to see that

$$\|T_N - T_{N+1}\| \leq C\left[\left(\frac{1}{N}\right)^{\frac{1}{n-2}} - \left(\frac{1}{(N+1)}\right)^{\frac{1}{n-2}}\right] \leq CN^{-1-\frac{1}{n-2}}.$$

We therefore conclude that $T = \lim_{N\to\infty} T_n$ is a compact operator on $C(\partial\Omega)$.

Next we apply the divergence theorem on $\Omega \mid B_\delta(P)$ for small positive δ's with $\delta \to 0^+$ to obtain

$$\int_{\partial\Omega} \frac{\partial}{\partial n_Q} R(P,Q)d\mathcal{H}^{n-1}(Q) = 1 \quad \text{if } P \in \Omega, \tag{6.18}$$

$$\int_{\partial\Omega} K(P,Q)d\mathcal{H}^{n-1}(Q) = \frac{1}{2} \quad \text{if } P \in \partial\Omega. \tag{6.19}$$

Let $P_0 \in \partial\Omega$ and $P \in \Omega$ such that $P \to P_0$. We want to verify that

$$\mathcal{D}f(P) \Rightarrow \tfrac{1}{2}f(P_0) + Tf(P_0). \tag{6.20}$$

Here we observe that $\int_{\partial\Omega} |\frac{\partial}{\partial n_Q} R(P,Q)| d\mathcal{H}^{n-1}(Q) \leq C < \infty$ for all $P \notin \partial\Omega$. Thus, in particular, $\|\mathcal{D}f\|_{L^\infty(\mathbb{R}^n|\partial\Omega)} \leq C\|f\|_{L^\infty(\partial\Omega)}$.

If $P_0 \notin$ support of f, then it is obvious that

$$\int_{\partial\Omega} \frac{\partial}{\partial n_Q} R(P,Q)f(Q)d\mathcal{H}^{n-1}(Q) \xrightarrow{P\to P_0} \int_{\partial\Omega} K(P_0,Q)f(Q)d\mathcal{H}^{n-1}(Q) = Tf(P_0).$$

If $P_0 \in$ support of f and $f(P_0) = 0$, then we let $\{f_k\} \subset C(\partial\Omega)$ such that

$$\|f - f_k\|_{L^\infty(\partial\Omega)} \xrightarrow{k\to\infty} 0,$$

and $P_0 \notin$ support of f_k for each k, $k = 1, 2, \ldots$. Then

$$\begin{aligned}
|\mathcal{D}f(P) - Tf(P)| &\leq |\mathcal{D}(f - f_k)(P)| + |T(f - f_k)|(P) \\
&\quad + |\mathcal{D}f_k(P) - Tf_k(P)| \\
&\leq C\|f - f_k\|_{L^\infty(\partial\Omega)} + \|T\|\|f - f_k\|_{L^\infty(\partial\Omega)} \\
&\quad + |\mathcal{D}f_k(P) - Tf_k(P)|.
\end{aligned}$$

We initially choose k large so that the first two terms on the right-hand side of the above inequality will be small. We then observe that for fixed k (large) as $P \to P_0$, the last term in the inequality also goes to 0.

To complete the proof it suffices to verify the case when $f \equiv 1$, for which the result is trivial. If we replace Ω by Ω^c, then all the other statements in Lemmas 6.16 and 6.17 follow. □

To conclude our consideration of double-layer potentials we need to show how to use them to solve the Dirichlet problem (6.10).

We begin with a $g \in C(\partial\Omega)$ and let $u(x) = \mathcal{D}g(x)$ for $x \in \Omega$. It is clear from our previous discussion that $\Delta u = 0$ in Ω and $u \in C(\overline{\Omega})$; moreover, $u|_{\partial\Omega} = (\frac{1}{2}I + T)g$. Therefore, we need to solve for g a given $f \in C(\partial\Omega)$, $f = (\frac{1}{2}I + T)g$. Since T is compact, $(\frac{1}{2}I + T)$ is obviously a $1:1$ map on $C(\partial\Omega)$; hence it is also an onto map from $C(\partial\Omega)$ to $C(\partial\Omega)$. This last statement follows from the Leray-Schauder fixed-point theorem, which we will examine in the next section.

Finally, we shall state without proof the results corresponding to those for single-layer potentials (6.16). All of the proofs are similar to those for Lemmas 6.16 and 6.17 above.

Once again we assume $\partial\Omega$ to be class C^2 and $f \in C(\partial\Omega)$.

LEMMA 6.18 *If* $f \in C(\partial\Omega)$, *then*

(i) $\mathcal{D}_+ S(f) = \operatorname{grad} S(f) \in C(\overline{\Omega_{\delta_0}})$ *and*

(ii) $\mathcal{D}_- S(f) = \operatorname{grad} S(f) \in C(\overline{\Omega^c_{\delta_0}})$.

Here $\overline{\Omega_{\delta_0}} = \{x \in \overline{\Omega} : \operatorname{dist}(x, \partial\Omega) \le \delta_0\}$ *for some small* $\delta_0 > 0$.

Let $K^*(P,Q) = K(Q,P)$ and define

$$T^* f(P) = \int_{\partial\Omega} K^*(P,Q) f(Q) d\mathcal{H}^{n-1}(Q), \qquad P \in \partial\Omega.$$

LEMMA 6.19 (Jump Relations for $\mathcal{D}S(f)$)

(i) $\mathcal{D}_+ S(f) = -\frac{1}{2}I + T^*$ *and*

(ii) $\mathcal{D}_- S(f) = \frac{1}{2}I + T^*$.

Single-layer potentials can be used to solve the Neumann problem

$$\begin{cases} \Delta u = 0 & \text{in } \Omega, \\ \frac{\partial u}{\partial n} = f & \text{on } \partial\Omega. \end{cases}$$

Layer potentials can be used to solve more general elliptic equations (and systems) with constant coefficients on smooth domains. This method can be further generalized to $C^{1,\alpha}$-domains for general elliptic equations of second order with C^α-coefficients (or general first-order elliptic systems with suitably smooth coefficients). The latter is often referred to as *ADN theory* due to Agmon, Douglis, and Nirenberg.*

6.6. Fixed-Point Theorems and Existence Results

The Brouwer fixed-point theorem asserts that a continuous mapping of a closed ball in $\mathbb{R}^n$ into itself has at least one fixed point. In this section we shall discuss a version of the fixed-point theorem in an infinite-dimensional Banach space due to Schauder and a special case of the Leray-Schauder theorem. As an application we shall discuss the existence of a class of quasi-linear elliptic equations for the

*Agmon, S.; Douglis, A.; Nirenberg, L. Estimates near the boundary for solutions of elliptic partial differential equations satisfying general boundary conditions. I, II. *Comm. Pure Appl. Math.* **12** (1959) 623–727; **17** (1964), 35–92.

Dirichlet problem. We shall end the section with a brief mention of the minimal surface equation.

THEOREM 6.20 (Schauder's Fixed-Point Theorem) *Let $\mathcal{G}$ be a compact, convex set in a Banach space X, and let T be a continuous mapping of $\mathcal{G}$ into itself. Then T has a fixed point; that is, for some $x \in \mathcal{G}$, $Tx = x$.*

PROOF: Let $k \in N$. Since $\mathcal{G}$ is compact, there is a finite set $\{x_1, x_2, \dots, x_n\}$ where $n = n(k)$ such that the balls $B_i = B_{i/k}(x_i)$, $i = 1, 2, \dots, n$, cover $\mathcal{G}$. Let $\mathcal{G}_k$ be the convex hull of $\{x_1, x_2, \dots, x_n\}$ and let $J_k : \mathcal{G} \to \mathcal{G}_k$ be defined by

$$J_k(x) = \frac{\sum_i \operatorname{dist}(x, \mathcal{G} - B_i) x_i}{\sum_i \operatorname{dist}(x, \mathcal{G} - B_i)}. \tag{6.21}$$

It is easy to see that J_k is continuous on $\mathcal{G}$; furthermore,

$$\|J_k(x) - x\| \le \frac{\sum_i \operatorname{dist}(x, \mathcal{G} - B_i)\|x - x_i\|}{\sum_i \operatorname{dist}(x, \mathcal{G} - B_i)} < \frac{1}{k}.$$

The mapping $J_k \circ T : \mathcal{G}_k \to \mathcal{G}_k$. Thus by the Brouwer fixed-point theorem, there exist $y_k \in \mathcal{G}_k$ such that $J_k \circ T(y_k) = y_k$, $k = 1, 2, \dots$. Since $\mathcal{G}$ is compact, one may assume, without loss of generality, that $y_k \to x \in \mathcal{G}$. Since

$$\|y_k - T(y_k)\| = \|J_k \circ T(y_k) - T(y_k)\| < \frac{1}{k},$$

and since T is continuous, we have

$$\lim_{k \to \infty} y_k = x = Tx \quad \text{for some } x \in \mathcal{G}.$$

□

COROLLARY 6.21 *Let $\mathcal{G}$ be a closed, convex set in a Banach space X. Suppose T is a map from $\mathcal{G}$ into $\mathcal{G}$ such that $T\mathcal{G}$ is precompact. Then T has a fixed point in $\mathcal{G}$.*

Note that a continuous mapping between two Banach spaces is called *compact* (or *completely continuous*) if the images of bounded sets are precompact; that is, their closures are compact.

THEOREM 6.22 (Leray-Schauder Theorem) *Let T be a compact mapping of a Banach space X into itself, and suppose there is a constant M such that*

$$\|x\| < M \tag{6.22}$$

for all $x \in X$ and $\sigma \in [0, 1]$ satifying $x = \sigma Tx$. Then T has a fixed point.

PROOF: Define a new mapping T^* by

$$\begin{cases} T^*x = Tx & \text{if } \|Tx\| \le M, \\ T^*x = M\frac{Tx}{\|Tx\|} & \text{if } \|Tx\| \ge M. \end{cases} \tag{6.23}$$

T^* is clearly a continous mapping of the closed ball $\overline{B_M}$ into $\overline{B_M} \subset X$ itself. Since $T(\overline{B_m})$ is precompact, the same is true for $T^*(\overline{B_M})$. Hence by Corollary 6.21, the mapping T^* has a fixed point x. We claim x is also a fixed point of T. Indeed, if $\|Tx\| \ge M$, then $x = T^*x = \frac{M}{\|Tx\|}Tx = \sigma Tx$ where $\sigma = \frac{M}{\|Tx\|} \in (0, 1]$. By hypothesis, we would have $\|x\| < M$. On the other hand, $\|x\| = \|T^*x\| = M$; a contradiction. Thus $\|Tx\| < M$ must be true and consequently $T^*x = x = Tx$. □

REMARK. Suppose T is a compact mapping of a Banach space X into itself. Then for some $\sigma \in (0,1]$, the map σT possesses a fixed point. Indeed, since $T(\overline{B_1})$ is compact in X, there is an $A \geq 1$ such that $\|Tx\| \leq A$ for all $x \in \overline{B_1}$. Thus the mapping σT with $\sigma = \frac{1}{A}$ maps $\overline{B_1}$ into itself and our conclusion follows. Also, note that if (6.22) is valid, then for any $\sigma \in [0,1]$, the mapping σT has a fixed point as well.

Next we shall describe a situation in which Theorem 6.22 can be applied. For $\beta \in (0,1)$, we consider the Banach space $X = C^{1,\beta}(\overline{\Omega})$ where Ω is a $C^{2,\alpha}$, bounded domain in $\mathbb{R}^n$. Let L be an operator given by

$$Lu = a^{ij}(x,u,\nabla u)u_{x_i x_j} + b(x,u,\nabla u). \tag{6.24}$$

We assume that L is elliptic in $\overline{\Omega}$; i.e., $(a^{ij}(x,\zeta,p))$ is positive definite for all $(x,\zeta,p) \in \overline{\Omega} \times \mathbb{R} \times \mathbb{R}^n$. We also assume, for some $\alpha \in (0,1)$, that $a^{ij}, b \in C^{\alpha}(\overline{\Omega} \times \mathbb{R} \times \mathbb{R}^n)$. Let $\phi \in C^{2,\alpha}(\partial\Omega)$. For all $v \in C^{1,\beta}(\overline{\Omega}) = X$, we let $u = Tv$ be the unique solution in $C^{2,\alpha\beta}(\overline{\Omega})$ of the linear Dirichlet problem

$$\begin{cases} a^{ij}(x,v,Dv)u_{x_i x_j} + b(x,v,Dv) = 0 & \text{in } \Omega, \\ u|_{\partial\Omega} = \phi & \text{on } \partial\Omega. \end{cases} \tag{6.25}$$

We note that the solvability of $Lu = 0$ in Ω with $u = \phi$ on $\partial\Omega$ in the space $C^{2,\alpha}(\overline{\Omega})$ is equivalent to the solvability of $Tu = u$ in X.

Let

$$L_\sigma u = a^{ij}(x,u,Du)u_{x_i x_j} + \sigma b(x,u,\nabla u). \tag{6.26}$$

Then $u = \sigma Tu$ in X is the same as $L_\sigma u = 0$ in Ω and $u = \sigma\phi$ on $\partial\Omega$. As a consequence of the Leray-Schauder theorem, we have the following:

THEOREM 6.23 *Let Ω, ϕ, and L be as above. If, for some $\beta > 0$, there is a constant M independent of u and σ such that every $C^{2,\alpha}(\overline{\Omega})$-solution of the Dirichlet problem*

$$\begin{cases} L_\sigma u = 0 & \text{in } \Omega, \\ u = \sigma\phi & \text{on } \partial\Omega, \end{cases} \tag{6.27}$$

satisfies

$$\|u\|_{C^{1,\beta}(\overline{\Omega})} < M, \tag{6.28}$$

then it follows that the Dirichlet problem $Lu = 0$ in Ω with $u = \phi$ on $\partial\Omega$ is solvable in $C^{2,\alpha}(\overline{\Omega})$.

PROOF: From the preceding discussion, it suffices to verify that T is continuous and compact. Again, this is simply a consequence of the Schauder estimates. We note that $C^{2,\alpha\beta}(\overline{\Omega})$ is precompact in $C^{1,\beta}(\overline{\Omega})$. □

We shall briefly mention how assumptions in Theorem 6.23 can be verified for the minimal surface equation. For simplicity, we consider the case where Ω is a uniformly convex, $C^{2,\alpha}$-bounded domain in $\mathbb{R}^n$ and the following Dirichlet problem:

$$\begin{cases} \operatorname{div}\left(\frac{\nabla u}{\sqrt{1+|\nabla u|^2}}\right) = 0 & \text{in } \Omega, \\ u|_{\partial\Omega} = \phi & \text{on } \partial\Omega. \end{cases} \tag{6.29}$$

We assume that $\phi \in C^{2,\alpha}(\partial\Omega)$.

Suppose u is a $C^{2,\alpha}$-solution of (6.28); then the maximum principle implies that

$$\|u\|_{L^\infty(\overline{\Omega})} \leq \|\phi\|_{L^\infty(\partial\Omega)} \equiv C_0 < \infty. \tag{6.30}$$

Next, by the uniform convexity of $\partial\Omega$ and the $C^{2,\alpha}$-regularity of ϕ, we can check that, for every $x_0 \in \partial\Omega$, there exist linear functions $\ell^{\pm}_{x_0}(x)$ such that

$$\ell^{\pm}_{x_0}(x_0) = \phi(x_0) \quad \text{and} \quad \ell^{-}_{x_0}(x) \leq \phi(x) \leq \ell^{+}_{x_0}(x)$$

for all $x \in \partial\Omega$. Since linear functions are solutions of $\operatorname{div}(\nabla u/\sqrt{1+|\nabla u|^2}) = 0$ in Ω, from the maximum principle we conclude that

$$\ell^{-}_{x_0}(x) \leq u(x) \leq \ell^{+}_{x_0}(x), \quad x \in \overline{\Omega}; \tag{6.31}$$

in particular, $|\nabla u(x_0)| \leq \max|\nabla \ell^{\pm}_{x_0}(x_0)| \equiv C_1 < \infty$.

On the other hand, if u is a $C^{2,\alpha}$-solution of (6.28), then $u_\alpha = \frac{\partial}{\partial x_\alpha} u$, $\alpha = 1, 2, \ldots, n$, satisfies

$$\frac{\partial}{\partial x_i}(F_{P_i P_j}(Du) u_{a_j}) = 0. \tag{6.32}$$

Here $F(Du) = \sqrt{1+|\nabla u|^2}$, hence $(F_{P_i P_j}(Du)) > 0$. Thus u_α satisfies the maximum principle. Therefore we have

$$\|\nabla u\|_{L^\infty(\Omega)} \leq \|\nabla u\|_{L^\infty(\partial\Omega)} \leq C_1 < \infty. \tag{6.33}$$

From (6.29), (6.32), and (6.33), we further deduce that

$$\|\nabla u\|_{C^\beta(\overline{\Omega})} \leq C(C_0, C_1, C_2) < \infty \tag{6.34}$$

where $C_2 = \|\phi\|_{C^{2,\alpha}(\partial\Omega)}$. This follows from De Giorgi–Moser theory.

We rewrite $\operatorname{div}(\nabla u/\sqrt{1+|\nabla u|^2}) = 0$ as

$$\Delta u - \frac{u_i u_j}{1+|\nabla u|^2} u_{ij} = 0 \tag{6.35}$$

and combine (6.35) with (6.34) and the Schauder estimates to obtain

$$\|u\|_{C^{2,\beta}(\overline{\Omega})} \leq C(C_0, C_1, C_2, \Omega) \tag{6.36}$$

where we may assume that $0 < \beta \leq \alpha$.

Bibliography

[1] Berestycki, H., Nirenberg, L., and Varadhan, S. R. S. The principle eigenvalue and maximum principle for second-order elliptic operators in general domains. *Comm. Pure Appl. Math.* 47: 47–92, 1994.

[2] Caffarelli, L. A. Interior a priori estimates for solutions of fully nonlinear equations. *Ann. of Math.* 130: 189–213, 1989.

[3] Caffarelli, L. A. Interior $W^{2,p}$ estimates for solutions of the Monge-Ampère equation. *Ann. of Math.* 131: 135–150, 1990.

[4] Caffarelli, L. A., and Cabré, X. *Fully nonlinear elliptic equations.* AMS Colloquium Publications, 43. American Mathematical Society, Providence, R.I., 1995.

[5] Giaquinta, M. *Multiple integrals in the calculus of variations and nonlinear elliptic systems.* Annals of Mathematics Studies, 105. Princeton University Press, Princeton, N.J., 1983.

[6] Gidas, B., Ni, W. M., and Nirenberg, L. Symmetry and related properties via the maximum principle. *Comm. Math. Phys.* 68: 209–243, 1979.

[7] Gilbarg, D., and Serrin, J. On isolated singularities of solutions of second order elliptic differential equations. *J. Analyse Math.* 4: 309–340, 1955/56.

[8] Gilbarg, D., and Trudinger, N. S. *Elliptic partial differential equations of second order.* 2nd ed. Grundlehren der mathematischen Wissenschaften, 224. Springer, Berlin–New York, 1983.

[9] John, F. *Partial differential equations.* Reprint of the 4th ed. Applied Mathematical Sciences, 1, Springer, New York, 1991.

[10] John, F., and Nirenberg, L. On functions of bounded mean oscillation. *Comm. Pure Appl. Math.* 14: 415–426, 1961.

[11] Krylov, N. V. *Lectures on elliptic and parabolic partial differential equations in Hölder spaces.* Graduate Studies in Mathematics, 12. American Mathematical Society, Providence, R.I., 1996.

[12] Littman, W., Stampacchia, G., and Weinberger, H. F. Regular points for elliptic equations with discontinuous coefficients. *Ann. Scuola Norm. Sup. Pisa (3)* 17: 43–77, 1963.

[13] Protter, M. H., and Weinberger, H. F. *Maximum principles in differential equations.* Prentice-Hall, Englewood Cliffs, N.J., 1967.

[14] Serrin, J. A symmetry theorem in potential theory. *Arch. Rational Mech. Anal.* 43: 304–318, 1971.

[15] Stein, E. M. *Singular integrals and differentiability properties of functions.* Princeton Mathematical Series, 30. Princeton University Press, Princeton, N.J., 1970.

Titles in This Series